International Conference on Marine Industry 4.0

Rotterdam, The Netherlands
5 November 2019

ISBN: 978-1-7138-0869-5

Printed from e-media with permission by:

Curran Associates, Inc.
57 Morehouse Lane
Red Hook, NY 12571

Some format issues inherent in the e-media version may also appear in this print version.

The Royal Institution of Naval Architects
8-9 Northumberland Street
London, WC2N 5DA
United Kingdom

Phone: 020 7235 4622
Fax: 020 7259 5912

publications@rina.org.uk

Additional copies of this publication are available from:

Curran Associates, Inc.
57 Morehouse Lane
Red Hook, NY 12571 USA
Phone: 845-758-0400
Fax: 845-758-2633
Email: curran@proceedings.com
Web: www.proceedings.com

International Conference on MARINE INDUSTRY 4.0

5th November 2019, Rotterdam, the Netherlands

PAPERS

Sponsored by:

CDPStudio

THE ROYAL INSTITUTION OF NAVAL ARCHITECTS
8-9 Northumberland Street, London, WC2N 5DA, Telephone: +44 (0)20 7235 4622

The Royal Institution of
Naval Architects

INTERNATIONAL CONFERENCE

Marine Industry 4.0

5th November 2019, Rotterdam, the Netherlands

Papers published in these conference Proceedings are peer reviewed.

THE ROYAL INSTITUTION OF NAVAL ARCHITECTS
8-9 Northumberland Street
London WC2N 5DA

Telephone: 020 7235 4622
Fax: 020 7259 5912

CONTENTS

Practical Guide for Adopting Industry 4.0 in CAD Systems 1
J. A. Muñoz and R. Pérez Fernández, SENER, Spain

Living on the Edge - a New Paradigm for Remote Monitoring of Vessels and Marine Equipment 9
K. Rau, Flicq Inc, USA

One Sea - on the road towards an autonomous maritime transport system 11
P. Haikkola and J. Merenluoto, One Sea / DIMECC, Finland

Aero-triangulation and photogrammetry in sensible infrastructures – how the industry benefits from low-cost 3d modelling conducted by unmanned aerial vehicles (UAV) 19
M. Stein, Hapag Lloyd Centre of Shipping and Global Logistics, Germany

The Challenges and Opportunities for Developing Digital Twins to support Marine Operations 31
S. Willmore, G. Mercer, J. Rigby, BMT, UK

Classifying bow entry events of wave piercing catamarans in random waves using unsupervised and supervised techniques 39
B. Shabani, J. Lavroff, D. S. Holloway, University of Tasmania, Australia
S. Penev, UNSW Sydney, Australia
D. Dessi, Italian National Research Council, Italy
G. Thomas, University College London, UK

Future Workforce and Training in the Fourth Maritime Revolution 55
R. Westgarth and J. Rigby, BMT, UK

IConIC – the autonomous health monitoring system* N/A
D. Garrity, STS Defence Ltd, UK

Digitalisation in the UK maritime sector: a stakeholders' pulse check 63
C. Gkerekos, G. Theotokatos, L. M. Bujorianu, E. Boulougouris and D. Vassalos, University of Strathclyde, Glasgow, UK
B. Carballedo and S. McCluskey, HSSMI, Stratford, UK

***Unavailable at time of going to press**

PRACTICAL GUIDE FOR ADOPTING INDUSTRY 4.0 IN CAD SYSTEMS

J. A. Muñoz and **R. Pérez Fernández**, SENER, Spain

SUMMARY

Technologies are evolving faster than our ability to assimilate what we can do with them, but the potential is clear and the opportunity will be for those who identifies the right application of each technology.

In the information era, we are literally swimming in an ocean of structured and not structured data and thanks to the evolution in the communications technologies, all that data are available everywhere for everyone. But data is not information. It is necessary to have the capability to analyse, extract conclusions and learn from it. Technologies as *Big Data* (BD) and the *Artificial Intelligence* (AI) are crucial for this purpose, but the intention of the treatment matters.

Imagine how these technologies shall allow to engage the ship design by applying rules which will facilitate the design significantly, how the integration of the validation of the structural models by the *Classification Societies* will be linked directly by cloud applications. Imagine all the benefits of this two simple examples that can be implemented thanks to the potential of these technologies.

The concepts that are absolutely clear from now to the future in shipbuilding is the use of *Data Centric* model and the concept of *Digital Twin*, a real and effective synchronization between what we design, what we construct, by covering the complete life cycle of the product thanks to technologies like *IoT*.

It is important to understand how the new generations are immersed in a technological world in constant and rapid evolution. The way they interacts with this ecosystem will determine the way we should define the new rules of the CAD/CAM/CIM Systems.

This paper examines different selected solutions describing practical use cases in ship design phase as an example of what *IoT, BD* or *AI* will represent for ship design and shipbuilding in the near future.

NOMENCLATURE

AI	*Artificial Intelligence*
AR	*Augmented Reality*
CAD	*Computer Aided Design*
CAE	*Computer Aided Engineering*
CAM	*Computer Aided Manufacturing*
GPS	*Global Positioning System*
IoT	*Internet of Things*
RFID	*Radio-Frequency Identification*
RPA	*Robotic Process Automation*
VR	*Virtual Reality*

1. INTRODUCTION

Industry 4.0 has favoured the expansion of many technologies whose application boundaries are very diffuse. Although each technology can have very specific applications, when it is wanted to apply to *Computer Aided Design, Manufacturing & Engineering* (*CAD*/CAM/CAE) System, from now on referred just as *CAD*, they have to be implemented as a whole.

CAD tool is in the beginning of the design but it manages many data that must be taken in advance for the further stages of the lifecycle of the product. Augmented reality, virtual reality and mixed reality are closely related to the *Digital Twin* and interlaced with the *Big Data*, which are generated by *CAD* tools and all surrounding solutions, which applies some cloud/edge/fog computing to these data in a merged technology between finite state machines and *Artificial Intelligence* (AI) cognitive processes (*Muñoz & Perez, 2017*).

To perform in an agile manner all these integrations requires a network which support different connection ways to add specific devices, i.e. *Internet of Things* (*IoT*), which can access to the data, creating and modifying them, in a different layer which affects to the basic information layer created by the *CAD* System in the shipyard.

This network should be secured but at the same time open to allow distributed work, which must be tracked in such a way that records any modification of every working step done in an open, transparent, trusted and non-modifiable working method for all stakeholders involved in process, like: shipyard, engineering offices, classification society and ship owner.

Results of the design should be easy integrated with future building ways like *3D* printing, generating printing orders directly from the *CAD* model.

Shipbuilding phases involve design and manufacturing, but an integrated *Industry 4.0 CAD* System should also be involved in operation and maintenance phases, i.e. it must cover all the vessel lifecycle end to end, from design to decommission.

When a ship comes for a reparation, sometimes the model is not available for this operation of maintenance (*Muñoz et al., 2018*). Trying to model into the *CAD* System, the entire engine room or any other local from scanned model could be a nightmare. *AI tools* can be used for processing the resulted cloud of points and creating geometries that can be used as references in the *CAD* or even can create equivalent models, which can be converted with a minor user intervention, and lesser as the *AI* learns, in a *CAD* integrated and full modifiable design.

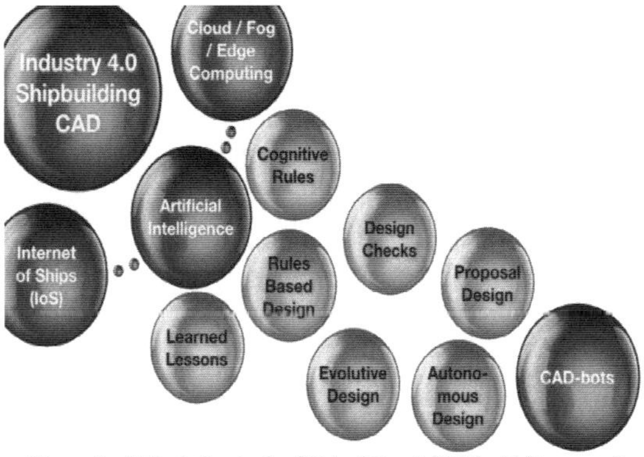

Figure 1: AI Evolution in the Shipbuilding CAD World (Benayas & Perez, 2019)

At the end, this is a brief sum up of the *Industry 4.0* technologies which can be applied to a *CAD* System (as illustrated in *figure 1*), included in, or as an integrated surrounding solution or as information generators for the evolutionary design process.

2. WHAT IS THE INDUSTRY 4.0

To be able to understand how *Industry 4.0* became today's buzzword, a look at its predecessors might give us a perspective on how this revolution in particular is different.

- **The First Industrial Revolution**. The industrial revolution in Britain came in to introduce machines into production by the end of the 18th century (1760-1840). This included going from manual production to the use of steam-powered engines and water as a source of power. This helped agriculture greatly and the term factory became a little popular. One of the industries that benefited a lot from such changes is the textile industry, and was the first to adopt such methods. It also constituted a huge part of the British economy at the time.
- **The Second Industrial Revolution** dates between 1870 and 1914 (although some of its characteristics date back to the 1850) and introduced pre-existing systems such as telegraphs and railroads into industries. Perhaps the defining characteristic of that period was the introduction of mass production as a primary means to production in general. The electrification of factories contributed hugely to production rates. The mass production of steel helped introduce railways into the system, which consequently contributed to mass production. Innovations in chemistry, such as the invention of the synthetic dye, also mark such period as chemistry was in a rather primitive state then. However, such revolutionary approaches to industry were put to an end with the start of *World War I*. Mass production, of course, was not put to an end, but only developments within the same context were made and none of which can be called industrial revolutions.
- **The Third Industrial Revolution** is much more familiar to us than the rest as most people living today are familiar with industries leaning on digital technologies in production. However, the third industrial revolution is dated between 1950 and 1970. It is often referred to as the *Digital Revolution*, and came about the change from analogical and mechanical systems to digital ones. Others call it the *Information Age* too. The third revolution was, and still is, a direct result of the huge development in computers and information and communication technology.
- **The Fourth Industrial Revolution** takes the automation of manufacturing processes to a new level by introducing customized and flexible mass production technologies. This means that machines will operate independently, or cooperate with humans in creating a customer-oriented production field that constantly works on maintaining itself. The machine rather becomes an independent entity that is able to collect data, analyse it, and advise upon it. This becomes possible by introducing self-optimization, self-cognition, and self-customization into the industry. Manufacturers and operators will communicate with computers rather than operate them.

Industry 4.0 can be considered as the set of technologies that applied to industry in general will allow the connection of machines, products and services to gather and analyse data enabling flexible, faster and efficient processes to produce cheaper and with high quality. The most relevant technologies considered in this scope are *IoT*, *cloud computing*, *cybersecurity*, *augmented* and *virtual reality*, *additive manufacturing*, *autonomous robots*, *simulation* and *blockchain*.

But if there is any technology that supports the majority, this is artificial intelligence *(AI)*. *AI* will probably be the most disruptive technology in this revolution. However, what is *AI* actually? It can be said that there is no single origin of the concept known as *AI* and therefore there is no consensus to define that concept, but to be able to understand any of the definitions that apply, it is convenient to know at least briefly the most relevant facts and some of the milestones in its history.

It can be considered that *AI* was born as a philosophical study on human intelligence based on the concern of man to imitate the behaviour of other beings with capabilities beyond the reach of human beings (such as flying or diving), reaching the point of trying to imitate itself. In this sense, it can be said that *AI* is the search to imitate human intelligence. It is clear that it has not yet been completely achieved, but it is also increasingly true that we are closer. The first man who became aware of his own existence and was able to think, surely wondered how his thought would work and would conclude the idea superior creator, an intelligent being capable of creating another one. In this sense, the idea of a virtual design for intelligence is as old as human thought.

In 1920, the Czech writer Karel Capek published a science fiction stage play called *Rossum's Universal Robots*. The play is about a company that builds artificial organic humans in order to lighten the workload of other people.

Although in the play these artificial men are called robots, they have more to do with the modern concept of android or clone. They are creatures that can be passed as humans and have the gift of being able to think.
The English mathematician Turing publishes an article entitled *Computing Machinery and Intelligence* that opens the doors to *AI* (*Turing, 1950*). The article itself began with the simple question: *Can machines think?*.
Later Turing proposed a method to evaluate if the machines can think, which got to know itself like the Turing test. The test, or *Imitation game*, as it was called in the document, was presented as a simple test to judge if machines could think.
1956 Dartmouth conference convened by McCarthy and where the term *AI* was coined. The conference was attended by researchers from *Carnegie Mellon University* and *IBM*, including: Minsky, Newell and Simon. In this conference extremely optimistic forecasts for the next ten years were made that were never fulfilled, which caused the almost total abandonment of the investigations during fifteen years, known today as *AI* winter (*Aggarwal, 2018*).
In the other side, the concept *IoT* was born in the research centre *Auto ID* in *MIT* abound 1999 in research for realizing the object *identification using radio frequency* (RFID) which is the main idea of Kevin Ashton. The idea of the research was to identify all the objects and all the people in some way permitting some of the objects could exchange data with others through internet. The object identification was the prerequisite of *IoT*.

3. ADAPTATION OF THE NAVAL INDUSTRY TO THE NEW TECHNOLOGICAL CHALLENGES

As Stuart Taylor declared in (*Taylor, 2016*): "The platform is the key to success. The "things" will get increasingly inexpensive, applications will multiply, and connectivity will cost pennies. The real value will be created in the horizontal platform that ties it all together, i.e. the new OS."
In an article published in the Harvard Business Review, Michel E. Porter and James E. Heppelmann in (*Porter et al., 2015*), described the new technology stack: "Smart, connected products require companies to build and support an entirely new technology infrastructure. This "technology stack" is made up of multiple layers, including new product hardware, embedded software, connectivity, a product cloud consisting of software running on remote servers, a suite of security tools, a gateway for external information sources, and integration with enterprise business systems". If a company wants enter in this business must have products that connects with the products of other companies. So, the necessity of having an opened product is increasing.
From our point of view, the mentioned platform it is not exactly an OS, but a set of interconnected applications, which share the same data about the product. Data that have been appearing since the early design phases and feed that single database, not only with the design attributes but also with the desirable values for optimal performance. Around this database will appear the applications that control the life cycle of the components, the sensors that collect the information of operation or operational life of the same objects and devices.
The key is that all devices, sensors, components are not connected directly to the *IoT*, but connected to a management core that controls the data to interchange to *IoT* for those devices allowed to enter in that network. In this way, it is possible to take advance of the fact that all the existing components in the ship have its respective model created, incorporated, and defined from the CAD system. We could say that CAD is the birthplace of the elements of the ship. Therefore, it is the CAD system that knows them and from where they can be assigned the mission and the level of participation in that world of *IoT*.
Future smart ships must be connected to *IoT* or they will not be smart. The connection of smart devices within a smart ship must be human controlled. The control should start from the design tools because they control the shipbuilding process from the early stages of the design up to the final production. The set of design tools, product lifecycle management and device must be inter connected among them and will be the platform for the smart ships connected to *IoT*. The information shared in the scope of the *IoT* must be managed by the human along the whole lifecycle of the ship, starting from the beginning of the initial design. This need requires the CAD tools to be prepared with specific characteristics to handle that information. This new ecosystem, opened, incorporating the new trend of technology but adapted to the specific environment of shipbuilding will be the Internet of Ships.

4. ARTIFICIAL INTELLIGENCE

Among the reasons that recommend the appliance of AI in marine projects, most of them are common to other industries. Nevertheless, there are some reason more specific of shipbuilding industry, and whose impact is significant.

- More complex projects. Ships, and therefore marine projects, are becoming more complex, with more demanding technical specifications, more strict national and international regulations, and with tighter budgets.
- Shorter delivery times. At the same time projects are more complex, the competitiveness obliges to shorten delivery times, and this means that all phases are compressed, but without producing any reduction in the quality of the final product.
- Non-expert workers. This fact, common to all industries, is sharper in the marine sector, as shipyards and technical offices are obliged to involve into the development of projects new staff, young or with no previous experience in ship design and shipbuilding. This implies the adoption of not optimized solutions, or of solutions out of marine practise, with negative impact with consequences not only in the construction, but also in the operation of the ships.
- More complex tools. Tools used in design, construction and operation of ships are increasingly complex trying to provide solutions to the objective

of shorten delivery times at reduced cost. The use of such a tools require knowledge that is difficult and takes time to get; time that does not exist according to project timing. Therefore, it is very convenient to count with virtual assistants that guide the users and allow them not only to learn quickly, but also to solve problems without spending time in their analysis.

- Loss of know-how. Cyclic evolution of shipbuilding industry forced shipyards and technical offices to dispense with significant staff. These leaves rely mainly in elderly people that have the most relevant experience and expertise. The problem is not only in the loss of experience, but also the lack of transmitting the know-how to younger generations.

All these reasons, and other ones not explicitly mentioned here, mean important risks in marine projects, including the design phase, that can be minimised with the use of AI techniques.

5. DIGITAL TWIN

Everybody speaks about the digital transformation but is not always clearly understood what it means, it may be clarify by describing the three main pillars where the scope of this transformation is based on:

- Digital services. The digital services includes the generation of a digital twin of the ship, with all the relevant information. This mock-up is going to be of high value, not only for the Classification Societies, but also for the shipyards during the phase of construction, and for the ship-owners along the exploitation phase.
- Customer Experience. Is about to improve the Customer experience by means of the digital certification and with a specific set of applications that will facilitate the tasks of inspection and maintenance accordingly with the Classification Society Rules.
- Operations. In this case, the enhancement shall be focus in the improvement of the effectiveness and efficiency of the interventions on board the ships and in the internal procedures of the Classification Societies.

The cybersecurity (and particularly the software piracy), due the technological advances that has been enhance the automatization of the ships and because are day by day more dependant of the communications, together with the new technologies based on software (dynamic positioning, consumption optimization, electronic systems, gas emission reducers, etc.), are making our hardware/software systems potentially one of the weakest point for the international navigation, therefore it must be enhance and secured. Both the digital twin and its fundamental derived benefit, the digital services, had been demanded to revaluate aspects such as cybersecurity, piracy control, accidents prevention, performance of the software or systems, etc.

Inside the concept of cybersecurity, the *Classification Societies* are defining a new classification notations, such is the *SYS-COM* from the *Bureau Veritas* (*Pancorbo, 2017*), that is assigned to those ships that will have the incremented reliability thanks to the communication services between the ships and the land base, in such a way that will allow the monitoring analysis as well as the resolution of issues remotely augmented the reliability of the ship and their operations meanwhile are using these automatization systems, for support and maintenance, as it is presented in figure 2.

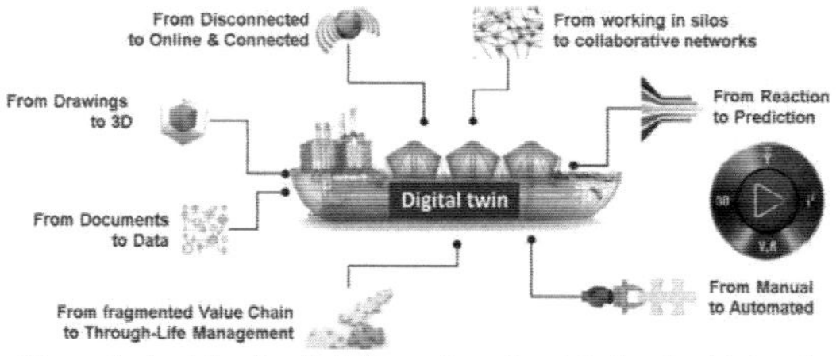

Figure 2: Applying the digital transformation into the Asset Integrity Management (Pancorbo, 2017)

Other tools that had already been in use are those oriented to help to the customer experience of the ship-owner:

- Mobile applications. Specific applications developed for ship-owners that can perform the tracking of the ship, handling its certificates, as well as classification services, etc.
- Specific cargo tools. Are those where it is possible to verify that our ship is ready to transport a particular cargo. Responding to specific needs from the shipping Company to verify, in the short term, the viability to transport a particular cargo that is not in the originally approved cargo list.
- E-Commerce. The e-commerce started as a web tool where the ship-owner can request the certification of the Ship drawings, at the same time that obtaining an on-line Price for it, having at the same time the possibility to pay with the credit card. It is specially oriented to improve ship-owners experience, above all by taking into account the financial advantages of reducing the time to get the already mentioned Class Certification.

6. VIRTUAL REALITY AND AUGMENTED REALITY

Shipbuilding industry is affected by a global and extremely competitive environment. All processes and systems have to be adapted to this challenging scenario, making special efforts in innovation and applying the most advanced technology available in the market. The Virtual Reality application to shipbuilding is not new, but it is more extended now thanks to the important improvements in software and hardware. From a user point of view, it is possible now to find a wide range of solutions to meet the most demanding requirements, to achieve measurable results, in terms of efficiency and costs.

The ship 3D model is the core to all tasks related to ship design and manufacturing.

It is the starting point of a series of thousands of tasks that are absolutely based on the coherence of the model. As an added value, CAD suppliers develop smart tools to use this model in realistic navigations in Virtual Reality environments with multiple purposes. Some of the most important uses of Virtual Reality in shipbuilding are described in the following paragraphs.

6.1. Virtual Reality during the ship design stages

Maybe the engineering department of a shipyard is where is more extended the use of this kind of solutions. And production department is maybe where it should be extended because is where there are the most costly errors. They are not usually working with a Virtual Reality solution but now it is starting to be clear that it is very useful to check the model, to avoid errors and inconsistencies but also to improve the tasks of production. So, the first and most important use of Virtual Reality is for avoiding errors. The possibility of viewing the ship 3D model as much realistic as possible is really effective to find out errors, in all stages of the design. From early design stages, to manufacturing and production phases, it is possible to use it to check all the elements, to see inconsistencies, to prevent interferences and collisions, to query about properties and attributes and to study different design alternatives and changes dynamically. In addition, the user and model interaction has become much more realistic with the implementation of tracking devices which give the sense of being inside the model, walking and moving on it, touching it.

Viewers allow an easy navigation and a fast-movement. Different modes of visualization, search and query options, inclusion of annotations and measurement of distances are important and useful capabilities. Ergonomic aspects in the design can be studied and checked better having a Virtual Reality solution, with the possibility to include dummies. It is possible also the collision detection, to check interferences and the insertion of annotations.

Simulation is another important area of application of Virtual Reality in ship design, engineering and production. It is being applied in the study of many tasks with different purposes, from the study of escape routes to the simulation of dismantling for maintenance of equipment items. The help in the study of critical assembly tasks and how the surrounding elements are affected is relevant too.

No need to say that the use of Virtual Reality for the control of the design is very useful, just to see the progress of the project. And more, once it is being built, the comparison between the model and the real ship is necessary and Virtual Reality can help to find out errors that need to be solved as soon as possible eliminating further unnecessary costs.

On the other hand, it is very common that the ship design and engineering and also the manufacturing is divided in different blocks that are subcontracted to different design offices, subcontractors and shipyards. The shipyard receives all the information of the ship and needs to supervise that everything is correct. Having a Virtual Reality tool that allows the integration of all the information in a simple way makes possible to have a single model, which is much more effective than having many different models.

6.2. Virtual Reality for commercial uses

Benefits of using Virtual Reality for marketing and commercial purposes are evident. That is why Virtual Reality tools are having a great reception in sales and marketing activities. Thanks to the wide range of possibilities available in the market, from small and portable solutions to big and on-demand solutions, the present and future applications of Virtual Reality is higher than we can imagine.

From the marketing point of view, the possibility to present a ship 3D model with the higher level of detail in an immersive Virtual Reality experience adds incalculable value to the sales activity. Many shipyards are already taking advantage of this technology, not only in their facilities but also in fairs and exhibitions thanks to the portable solutions that will be described later.

6.3. Shipyard management

Shipyard management play the role of ensuring that the commitments made to the clients are fulfilled. The project needs to be accurate, with the highest quality level and in accordance with estimations to ensure customer satisfaction. To achieve it, it is very important to have a clear idea about the progress of the project and also about changes that can affect further requirements, schedules and costs. Having the opportunity to check the Virtual Reality model from early design stages is a great advantage and allows a friendly checking, quick evaluation of alternatives and fast decision-making. At the end, the great impact in cost reduction is possible thanks to the early error detection, being much more expensive the modifications in manufacturing and production stages.

6.4. Ship-owners management

The presentation of the progress of the ship project to ship-owners is another important use of Virtual Reality, adding the value of being very easy to conceive the project in a very realistic and intuitive way. This has become a great advantage in comparison with conventional presentations. In fact Virtual Reality can substitute the old scale ship models that are expensive and with short level of detail. The capabilities to present the information available in the Virtual Reality solutions are as big as the user´s ability, because now it is possible to provide realistic finishes and all the level of detail required. Ship-owners can also check and monitor the progress of the model and also to promote design alternatives less cost-effective than in further stages.

6.5. Naval shipbuilding management

Military shipbuilding industry is affected by the most demanding requirements both in the design and production stages. It is true that this kind of naval projects, both in submarines and in surface ships, are lengthy and complex although they have more human resources. But, at the end, the project needs to comply with the strictest rules and also with the budgets imposed by the ministries. This is not an easy task, and that is why all the help in the control and supervision of the process is really welcome. Is in this area of shipbuilding where Virtual Reality

solutions have been well received time ago. The application of the most advanced technologies usually starts in ambitious programmes with major innovations and budgets. That is the case of application of Virtual Reality in shipbuilding, because we cannot forget that having an appropriate Virtual Reality solution is costly, and that long-term military projects are those that a priori have a larger budget to afford them.
Other important use around the naval shipbuilding is in the side of the Army. The use of Virtual Reality is an advantage for the crew training. In submarines this can be particularly interesting for the lack of space. Simulation is the other important activity, since there is really difficult to enhance some kind of operations inside a submarine but also in a surface ship. The help of Virtual Reality to simulate operations and dismantling and maintenance tasks is unquestionable. Here there is a great area of improvement.

6.6. Areas of improvement

From another point of view, there are still some drawbacks to mention around the use of this kind of technology in shipbuilding. The most important one is the price, because although there are cheaper devices available in the market, the true is that the hardware is expensive, and it is necessary to add the costs of the software, the implementation, the conditioning of the room, the training, etc.
Another important drawback is that there is necessary to have a good ship 3D model of the ships. Sometimes this is not possible, because there are multiple CAD systems applied in the same project, with the difficult of having just one model of the ship with all the information. To avoid this situation it is much better to develop the whole project in the same CAD, or at least to have the necessary tools to integrate the information in a viewer compatible to all of them, which sometimes it is a very difficult task.
Finally, the ergonomic aspect of the solution is important too. The use of 3D glasses for a long time is annoying for the eyes, and the standing position looking to a screen without any light could be unpleasant if used too often.

7. CONCLUSIONS

After the review of all this technology, it is possible to have an idea of the magnitude and complexity of the changes that *Industry 4.0* is demanding.
In most cases, it is an unavoidable step that must be taken to remain competitive, but with the adoption of only a part of what is exposed; it could mean a too risky and complex step that should be studied carefully before taking any decision. Some of these technologies are in early stages for industrial implementation but it is important to be aware and analyse what can offer and what can be useful for each case.
Most of the different technologies exposed in this paper are well known since long, but the true potential came a few years ago with the improvements in the communications infrastructures; and that is why a concept like Digital Twin is something that will be achievable.
Now with *5G*, it is expected that *IoT* will grow in an exponential way and its true potential will arise, which will also means the *Digital Twin* impulse.
Other technologies like *VR* have reappeared mainly thanks to the Game Development Industry and the lower cost of technology derived from the evolution of flat screens and miniaturization among other factors. *VR* seems to have an important future for training and simulation purposes, but its implementation in companies has a lower interest compared with *AR*.
Contact lenses that works with *AR*, are science fiction now? *DARPA* is interested in a new wirelessly-connected contact lens recently unveiled in France. These contact lenses provides augmented vision assistance to users by relying visual information wirelessly.
There is another important piece of the puzzle: the security, veracity and trust in the data transactions between all the different participants. *Cloud + security* are two pieces of the game and without them, all this ecosystem will never work. Data transactions should be guaranteed and all the steps should be tracked from the first step to the last one of the process thanks to *blockchain*.
This data centric model will be complemented with *Big Data*, *Analytics* and *AI*:

- Extracting conclusions from that big amount of data acquired from *IoT* devices.
- Making possible a better predictive maintenance thanks to the capability to predict failures or detect problems in early stages as well as the consequences of those problems.
- Learning from manpower that works in our Industries by analysing the activities in order to process all that information and be able suggest better practices by detecting bottlenecks and possible solutions.

Repetitive tasks and dangerous activities that nowadays are carried out by labour force, can be replaced or supported by robots, which means an important change for the companies and the people that works there, but this transformation is unstoppable. New jobs will be created and new profiles will be necessary but those jobs that could be replaced totally or partially with robots will disappear. People should react to this new conditions and adapt.
The implementation of robots and autonomous vehicles in the Industry will also mean the presence of new tools that will improve working conditions for current workers, but it is necessary an adaptation to all these new technologies and sometimes it will not be easy. Shipbuilding is an Industry that could take advantage by using them, but depending on the particularities of each shipyard, this technology will be more or less easy to implement and even if it is feasible, the decision should be taken with perspective, because physical robots and autonomous vehicles needs special conditions to work. A different case is the use of *RPA*, which could be easier to implement and very useful if the processes are clearly defined and can be automated.

New materials and *3D* printing techniques will have an important role in the future of Shipbuilding as is already being demonstrated in other industries such as aeronautics and automotive but in shipbuilding there are interesting challenges, the use of *3D* printing on board.
In the middle of this ecosystem we will find a *3D model*, created mainly with *CAD* Tools and this means that the importance of *CAD* will be even more capital than today. The interface between the *3D model* and the rest of the *Industry 4.0* is the Digital Twin, the link between the reality and the virtual world, both living concurrently during the evolution of the construction and extending this world of possibilities to the entire lifecycle of the product. CAD tools, as an important part of all this environment, should also evolve to be easily linked to these technologies, but also it is important to be adapted to the new generation of users that demands a different kind of interfaces and workflow and this will require an important effort.
This big world is now open and Shipbuilding is starting in it, but the potential is clear and to remain competitive is mandatory to study how these technologies can improve the benefits by upgrading processes, resources, workflow, cooperation between stakeholders… the tools are here, now it is necessary to study and analyse our particular case. There is no global solution for all industries, not even a global solution for a particular industry, but digitalization is on hand and it is mature enough to start implementation and be part of our strategy for the future.

8. REFERENCES

[1] Aggarwal, A. (2018). *"The Birth of AI and the first AI Hype Cycle"*.

[2] Benayas-Ayuso, A. and Perez Fernandez, R. (2018). "Automated/controlled storage for an efficient MBOM process in the shipbuilding managing the *IoT* technology". SMART SHIP, 23 and 24 January.

[3] Benayas-Ayuso, A. and Pérez Fernández, R. (2019). *"What should shipbuilding expect from the CAD/CAM systems of the future?"*. The Naval Architect magazine. April. pp. 28-31.

[4] Capek, K. (2017). *"R.U.R. (Robots Universales Rossum)"*. Books Mablaz.

[5] Muñoz, J.A.; Pérez, R. (2017). *"CAD tools for designing smart ships in the world of the internet of things (IoT)"*, SMART SHIP 2017, 24 and 25 January 2017.

[6] Muñoz, J.A.; Pérez, R.; Gutierrez, J.R. (2018). *"Design Rules Evaluation through technologies of treatment of Big Data"*, 17th COMPIT, Pavone.

[7] Pancorbo, J.: *"Bureau Veritas: evolution towards a digital industry"*, Revista Ingeniería Naval. pp. 75-80, June, 2017.

[8] Porter, M.E. and Heppelmann, J.E. (2015). *"How Smart, Connected Products Are Transforming Companies"* Harvard Business Review, October.

[9] Taylor, S. (2016). *"10 Predictions for the Future of IoT"*. Big Data Quarterly, March.

[10] Turing, A.M. (1950). *"Computing Machinery and Intelligence"*. Mind 49, pp.433-460.

9. AUTHORS' BIOGRAPHIES

Rodrigo Perez holds a PhD in Marine Engineering. He is Head of the Naval Shipbuilding Area at SENER. He is also professor in several subjects at the Marine Engineering School of the Technical University of Madrid.

Jesus Ángel Muñoz holds the current position of FORAN System Product Manager at SENER Ingeniería y Sistemas S.A. He is responsible for the technical development guidance of the FORAN System. He is also professor of Applied Systems Engineering at the Marine Engineering School of the Technical University of Madrid.

LIVING ON THE EDGE – A NEW PARADIGM FOR REMOTE MONITORING OF VESSELS AND MARINE EQUIPMENT

K. Rau, Flicq Inc., USA

SUMMARY

Adoption of Internet of Things (IoT) technologies in the marine industry is accelerating and the applications result in a vast amount of data to be processed. By deploying edge computing, independent and autonomous decisions can be made by vessels and remote marine assets without relying on constant communication with servers on shore or in the cloud. This paper proposes a system architecture that enables condition monitoring of remote marine assets with higher performance, improved latency and reduced total cost of ownership.

1. INTRODUCTION

With digital transformation and Industry 4.0 pervading the marine ecosystem, vessels and other marine assets are increasingly equipped with retrofitted or in-built sensors that help monitor their operating conditions.

The first generation of the Industrial Internet of Things (IIoT) architecture was primarily based on analytics platforms that were deployed in the cloud. Over the past few years, an alternate approach has emerged and is finding its way into the marine market.

By taking advantage of a decentralized architecture [1][2], edge computing provides several advantages over conventional IIoT systems including higher performance, lower latency and lower total cost of ownership.

In this paper, I have proposed an architecture for edge computing in the marine industry context and evaluated the benefits based on it. Section I introduces the components of an edge computing architecture and Section II describes use cases that could benefit from such an architecture.

2. ARCHITECTURE

2.1 OVERVIEW

Edge computing provides high performance computing resources, storage and connectivity between edge devices and the cloud. While cloud computing requires massive amounts of data to be transferred between the edge devices and the cloud servers, edge computing eliminates the processing overhead by analysing data close to its source. One of the key benefits of edge computing is extremely low latency due to the proximity to the source of the data.

2.2 WORKFLOW

The edge computing architecture can be broken down into three steps as illustrated in Fig. 1 below.

2.2 (a) Data collection

Sensors and industrial gateways collect data from marine assets.

2.2 (b) Edge analytics

Industrial gateways deployed at the edge and, in the last few years, smart sensors capable of performing data analytics observe, process and analyse streaming data in real time.

2.2 (c) Data modelling and visualization

The edge devices transfer selected data sets to the cloud for further analysis, to train machine learning models and to display the data.

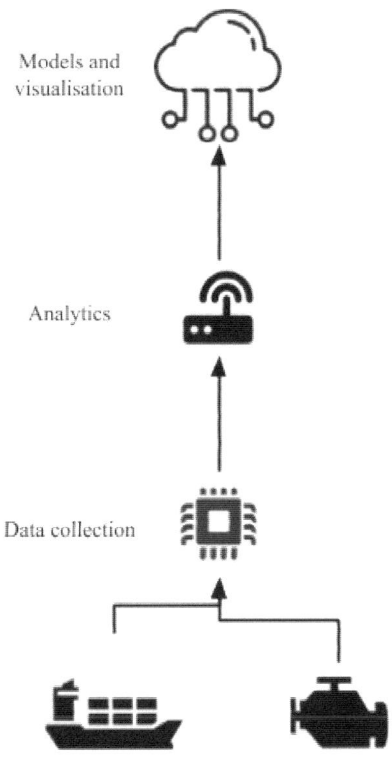

Figure 1: Edge computing architecture overview

2.3 IMPLEMENTATION

Data acquisition is a critical step in setting up an efficient edge computing system. Sensors (that measure vibration, motion, position, temperature, humidity, barometric pressure, etc.) are attached to marine/maritime equipment such as engines, alternators, pumps and compressors either using magnets or other means.

Sensors sample critical data periodically and transfer it to the edge gateway for processing. The gateway software stack analyses the data and defines a machine or equipment condition baseline [3]. All subsequent data sets are then compared with the baseline and the software determines if an anomaly has occurred. If an anomaly is detected, the gateway generates alerts and/or alarms that can be acted upon within milliseconds of detection. The inherent low latency of edge computing systems is one if its advantages compared to cloud based IoT systems.

The gateways require no additional infrastructure to function and are ideal for remote locations such as on-board vessels. Since a significant portion of the data is analysed on-board, edge computing systems obviate the need to transfer all the data to the cloud, resulting in significant bandwidth and total cost of ownership savings compared to cloud implementations.

For more compute intensive operations like Artificial Intelligence (AI) modelling and to provide an online dashboard for tracking, it is necessary for the gateways to send relevant information to the cloud.

3. APPLICATIONS

Remote condition monitoring [4] continues to be the most popular application for edge computing. In remote locations, where it is generally difficult to find ubiquitous wireless networks, edge computing can optimize bandwidth by processing data locally.

In addition, marine companies can take advantage of low latencies to implement predictive maintenance solutions. Edge computing is ideal for such applications due to short feedback loops between data analytics and modelling tools.

4. CONCLUSIONS

In this paper, an edge computing framework for marine and maritime applications has been presented. The generic workflow can be used to architect an edge computing solution and companies will be able to realize significant savings over prevailing cloud-based alternatives.

In the future, it will be worthwhile to monitor advancements in the areas of smart sensors and AI. Both technologies can enable additional applications at the edge and have the potential to push marine companies further down the path of digital transformation.

5. REFERENCES

1. Wu D, Greer MJ, Rosen DW, Schaefer D. Cloud manufacturing: strategic vision and state-of-the-art. J Manuf Syst 2013;32(4):564–79.
2. Zhang Z, Li X, Liu Y, Xie Y. Distributed Resource Environment: A Cloud-Based Design Knowledge Service Paradigm, Cloud-Based Design and Manufacturing (CBDM). Springer; 2014. p. 63–87
3. Aazam M, Huh E-N. Fog computing and smart gateway based communication for cloud of things, Proc. Future Internet of Things and Cloud (FiCloud). International Conference on, IEEE 2014:464–70.
4. Wright P, Dornfeld D, Ota N. Condition monitoring in end-milling using wireless sensor networks (WSNs). Trans NAMRI/SME 2008:36.

6. AUTHORS BIOGRAPHY

Karthik Rau holds the current position of CEO at Flicq Inc. based in Silicon Valley. He is responsible for setting the company's strategy and managing day-to-day operations. His previous experience includes Senior Director of Engineering at Atmel Corporation, where he led the design and development of IoT products. He previously held engineering and management positions at Intel Corporation and Gainspan Corporation.

ONE SEA – ON THE ROAD TOWARDS AN AUTONOMOUS MARITIME TRANSPORT SYSTEM

P. Haikkola and **J. Merenluoto,** One Sea / DIMECC, Finland

SUMMARY

Increased automation in the form of remote-controlled and partially autonomous operations are being tested around the world. This development is expected to create safer, more efficient, sustainable, and reliable maritime logistics chains that will enable more efficient ways of transport globally. Much of the digital transformation is happening in ecosystems. The maritime transformation and development are also collaborative efforts, as no single entity can make the transformation by themselves. One Sea is the leading industrial alliance in maritime autonomy that seeks to enable an autonomous maritime ecosystem by 2025. The alliance, founded in 2016, gathers together globally leading maritime actors, from equipment manufactures to ship owners and ICT companies. The paper introduces One Sea and its activities and the steps towards safe autonomous and remote-controlled maritime operations. It also addresses the recent achievements and practical examples of what One Sea partner companies have done in maritime autonomy.

NOMENCLATURE

AI – Artificial Intelligence
CAPEX – Capital Expenditure
ICT – Information and Communications Technology
IMO – International Maritime Organization
IoT – Internet of Things
IPR – Intellectual Property Rights
MASS – Maritime Autonomous Surface Ships
OPEX – Operating Expense
SRA – Strategic Research Agenda

1. INTRODUCTION

Digitalisation moves the value chains of traditional industry towards networks, where people, products, appliances and services as well as production systems and communities are seamlessly connected. Competition will shift from individual products and services towards value networks built around the needs of the customers. Disrupting business models are emerging in all industries and new players that did not exist few years ago are entering the market. Examples of new entrants in other industries include Tesla, Uber, Airbnb and Amazon to name a few. Maritime industry is following the technology transformation that sweeps across all industries.

Digitalisation opens opportunities for maritime industry as well, by transforming maritime and inland waterway transport in near future. It will be a powerful push towards new innovations. The early adopters will get clear benefits sometimes by being the first to enter the market, thus being able to lead the development of technologies and standards for the market as well as having increased opportunities to receive public support towards their first of a kind deployment. They will be able to exploit new business models and new skills.

Although the concept autonomous shipping is somewhat controversial, it is perceived by many the future of modern maritime traffic. Regarding the terminology for automated and connected solutions, there is still much confusion in the waterborne industries. There are no commonly agreed definitions including levels of autonomy, such as the other transport modes have, and different actors use otherwise different terminology. The IMO regulatory exercise talks about Maritime Autonomous Surface Ships (MASS), and that term is used at least on the maritime side. According to IMO, MASS is defined as a ship which, to a varying degree, can operate independent of human interaction. It needs to be emphasized that autonomous ships do not necessarily mean unmanned ships. Unmanned ships have high level of autonomy. However, manned ships can have high level of autonomy as well. Ships consist of multiple systems (propulsion, navigation, etc.), which can have different maturity regarding autonomy. The level of autonomy can also vary over time (part of voyage or ships lifecycle). The level of autonomy can be increasing automation, using various technologies. These include sensors, IoT, Big Data, smart algorithms, AI (Artificial Intelligence) et cetera.

The pursuit of safety, efficiency, sustainability and reliability are the key drivers for maritime digitalisation and autonomy. Safety comes first. Human errors cause 75-95% maritime accidents [1]. Autonomous technologies will reduce likelihood of accidents as well as the number of lost lives in maritime accidents. The second driver for the new technologies is cost. The economic benefit results from several factors and can reduce both CAPEX and OPEX. savings include for example lower fuel consumption and the reduced number of systems that need to be installed and maintained onboard. The third driver for the development is sustainability. Lower fuel consumption results in a smaller ecological footprint and more sustainable logistics. A biproduct of the development is increased reliability. The importance of system reliability increases as the level of autonomy increases in ships and the onboard crew decreases. Increased reliability highlights the need for standardization. Standardization enables better system validation and health management, increases the level of automation in the supply chain and supports transparency. Finally, it enables new types of businesses and business models.

Future ships of increased levels of autonomy operate in an ecosystem rather than in isolation. This requires redefining and assessing the roles and duties of the on-board and the crew ashore. The current shipping business is built on the collaboration between a large number of different actors in various roles. Maritime digitalisation and increased levels of autonomy will alter the existing roles and introduce completely new roles. Maritime autonomy will also make shipping a more integrated part of the transport and logistic chains. It will make logistics leaner by removing middlemen and in many instances putting the cargo owner in control of the supply chain. Autonomy can also change the entire system of transporting goods. The size of the transport units as well as ships can be expected to change when digitalisation will enable more individual transport solutions straight to customers.

One Sea seeks to enable autonomous transport system, which means increasing the level of autonomy on-board and interconnecting the maritime transport to other modes of transportation: road, rail and air. The maritime transport system comprises both physical infrastructure (ships, ports, fairways, freight and both terrestrial and satellite communication infrastructure), digital infrastructure (cloud services, data interfaces and platforms, connectivity and cybersecurity, AI), as well as services and new business models enabling the interoperable travel and transport chains. It includes addressing any regulatory, ethical and human factors as well. The transport system must be scalable on a global level. Successful outcome is a result of wide collaboration and mechanisms supporting the strategic and effective collaboration between public and private sectors as well as research.

2. BACKGROUND OF ONE SEA

2.1 PREPARATION

IMO's Maritime Safety Committee discussed automated ships as early as 1964 [2]. Autopilot has been used on ships for almost 100 years [3]. Condition-based monitoring systems have been offered to different ship systems at least since 1990ies. But it was not until the early part of 2010 that the waterborne industries really started taking automated waterborne traffic seriously. The solutions that are being built are mainly not built on current autopilots, but rather by utilizing Dynamic Positioning (DP) technologies, that have also been in use in some areas of the waterborne field already for some time, with the first commercial DP system being released in 1977 [4].

One Sea is a leading industrial alliance advancing maritime autonomy. The origin of One Sea is in strategy work completed by the Finnish marine industry and its stakeholders who drafted the Strategic Research Agenda 2025 of the Finnish Maritime Cluster, published on 1st November 2016, which outlines a roadmap for research in the industry [5]. The SRA work created an initial vision on autonomous vessels, where digitalisation becomes tangible in intelligent equipment and systems, as well as in digital services. The objective was to have the first autonomous vessels to sail within few years. The aim was to conduct a wide range of sea trials of digital solutions. The vision created in the SRA was translated into an ecosystem for autonomous marine transport, and DIMECC Ltd was selected to lead the ecosystem called One Sea.

DIMECC stands for Digital, Internet, Materials & Engineering Co-Creation. DIMECC is a multi-organisationally owned non-profit company. It has led the digital transformation and co-creation activities of manufacturing industries in Finland since 2008. DIMECC's network consists of 2.000+ R&D&I professionals, 400+ organizations, 69 shareholders and 10+ co-creation facilitators.

2.2 IMPLEMENTING THE ECOSYSTEM

The Ecosystem for Autonomous Marine Transport was started on 22 September 2016, as the Finnish Funding Agency for Innovation, Business Finland, granted the funding for the ecosystem. The ecosystem was initially co-funded by Business Finland and the partner companies. The continued activities in the future will be funded by the partner companies.

To understand the actions that One Sea needed to take, the ecosystem drafted a common development roadmap for the implementation of autonomous, partially manned and unmanned maritime transportation.

Although One Sea ecosystem originates from Finland it is an international initiative that connects some of global pioneers of maritime and ICT industries. One Sea soon turned towards international activities, rather than local Nordic or European activities, as its founding companies were all international. Maritime activities are global, and all efforts to increase automation need to be applied globally. Several global companies and partners have joined One Sea since its founding. One Se is a pre-competitive business-driven alliance bringing competitors and collaborators around the same table to work on a common agenda.

The ecosystem is a link between the key players in the maritime cluster, providers of digital solutions and the public sector. In practice, the ecosystem is creating the circumstances needed for efficient collaboration and coordinates development projects between different players to implement a common roadmap. One Sea is an open alliance welcoming industrial partners to join the ecosystem. The current list of ecosystem partners is listed in Table 1.

Table 1: One Sea partners.

Organisation	Description
ABB	ABB is a pioneering technology leader in electrification products, robotics and motion, industrial automation and power grids, serving customers in utilities, industry and transport & infrastructure globally. Continuing more than a 125-year history of innovation, ABB today is writing the future of industrial digitalisation and driving the Energy and Fourth Industrial Revolutions. ABB operates in more than 100 countries with about 135,000 employees.
Cargotec	Cargotec is a leading provider of cargo and load handling solutions with the goal of becoming the leader in intelligent cargo handling. Cargotec has three business areas: Kalmar, Hiab and MacGregor. Kalmar and Hiab provide solutions and services for cargo handling in ports, terminals and on-road whereas MacGregor offers engineering solutions and services for handling marine cargoes and offshore loads.
Ericsson	Ericsson is a world leader in communications technology and services. Their organization consists of more than 111,000 experts who provide customers in 180 countries with innovative solutions and services. Together they are building a more connected future where anyone and any industry is empowered to reach their full potential.
Finnpilot Pilotage	Finnpilot Pilotage's mission is to ensure the safe and smooth flow of maritime traffic. Their work reduces environmental accidents. The company employs over 300 pilotage professionals, who work 24/7 to carry out approximately 25,000 pilotage assignments each year. The company pilots 32% of the ships entering all Finland's ports from more than one thousand different fairways, working tirelessly to provide the safest, highest quality service to their customers.
Inmarsat	Inmarsat is the leading provider of global, mobile satellite communications services, creating fast, reliable connectivity on land, at sea and in the air.
Kongsberg Maritime	Kongsberg is an international knowledge-based group delivering high-technology systems and solutions. The company works with clients within the oil and gas industry, merchant marine, defence and aerospace. Kongsberg has 7 000 employees located in more than 25 countries, creating a total revenue of NOK 14,5 billion in 2017.
MTI (Monohakobi Technology Institute)	MTI Co., Ltd., a strategic subsidiary of Nippon Yusen Kabushiki Kaisha (NYK), takes a key role in the technological research and development field. MTI was established on April 1, 2004, to utilize technologies and the human resources of monohakobi in ways that realize customers' satisfaction and contribute to the global environment.
Tieto	Leading Nordic software and services company. Tieto aims to capture the significant opportunities of the data-driven world and turn them into lifelong value for people, business and society. They aim to be customers' first choice for business renewal by combining their software and services capabilities with a strong drive for co-innovation and ecosystems.
Wärtsilä Marine Solutions	Wärtsilä Marine Solutions enhances the business of its marine and oil & gas industry customers by providing innovative products and integrated solutions that are safe, environmentally sustainable, efficient, flexible, and economically sound.
Finnish Marine Industries	Finnish Marine Industries is a co-operation forum for high-technology maritime solution providers, leading marine equipment manufacturers, turn-key suppliers, designers, software and system providers as well as shipbuilding, ship repair and offshore yards.
Finnish Port Association	Finnish Port Association (FPA) brings the position of Finnish port authorities to decision making in Finland and the EU. FPA represents ports in cargo as well as passenger traffic. The membership of FPA consists of 30 sea and inland ports. The members of FPA are both general as well as industrial ports.
The Finnish Shipowners' Association	The Finnish Shipowners' Association is a trade association with an interest in industrial and labour market policies. Its members include 25 Finnish shipping companies and 105 ships.
The Royal Institution of Naval Architects	The Royal Institution of Naval Architects (RINA), is an internationally renowned professional institution whose members are involved at all levels in the design, construction, maintenance and operation of marine vessels and structures. Members of RINA are widely represented in industry, universities and colleges, and maritime organisations in

	over ninety countries.
Shipbrokers Finland	Shipbrokers Finland is the leading organisation in the field of shipbroking in Finland. Its aim is to promote the operational conditions for its member companies and the Finnish commercial navigation in co-operation with domestic and international stake-holders. The Association was founded in 1920, and its membership comprises 40 companies engaged in shipping services and logistics, which are either based or operating in Finland.
Business Finland	Business Finland is the most important publicly funded expert organization for financing research, development and innovation in Finland. They boost wide-ranging innovation activities in research communities, industry and service sectors.

3. ECOSYSTEM ACTIVITIES

One Sea has two primary functions. First, to enable the means for autonomous maritime operations and second, to communicate the message of maritime autonomy and its progress. The ecosystem's activities are centred around these functions. Figure 1 shows the ecosystem activities, which are discussed in greater detail in the next sections.

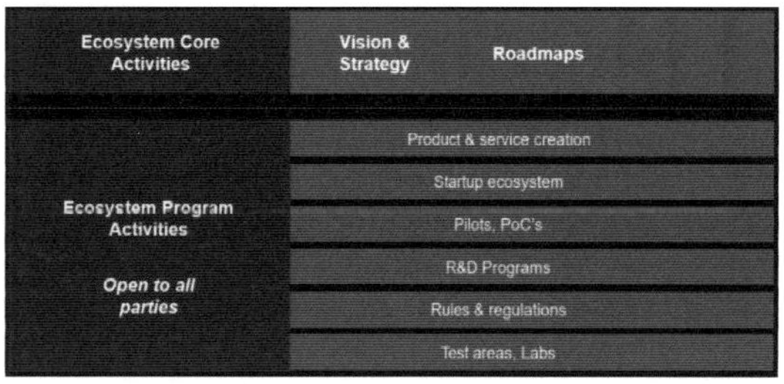

Figure 1: One Sea Ecosystem activities.

3.1 ROADMAP

The One Sea core activities include creation and steering the 2025 vision of the ecosystem and establishing the strategy to pursue the achievement of the vision. Another important core activity has been the creation of roadmaps towards 2025. The roadmap (Figure 2) published in the spring 2017 highlights the perceived technology development, milestones and themes on the way towards commercial autonomous shipping, as seen in early 2017.

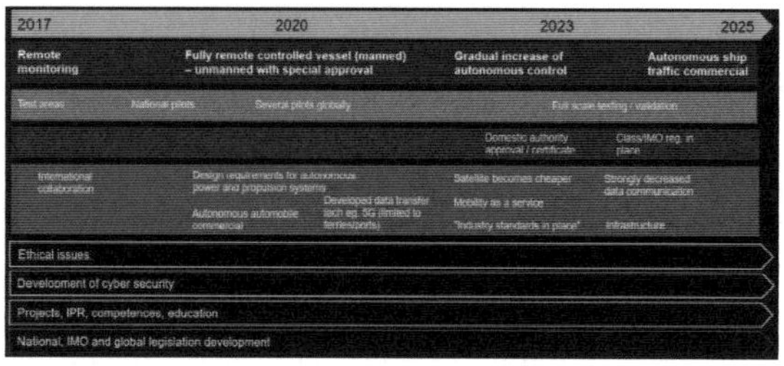

Figure 2: Roadmap towards 2025.

Top part of the roadmap envisions that focus will be initially on remote operations, which will gradually move towards more autonomous operations. The need for testing and validation is emphasized on the green background. It is acknowledged that testing is needed both to prove the technology, processes and procedures as well as to assure the regulators and class societies that the solutions are mature and safe. It will also assist in building the social acceptance for maritime autonomy. The path of regulation is on the red background. It is understood that the current regulation needs to be reworked to accommodate autonomous ships. The regulatory work takes time, hence the red background. The light blue area in the roadmap describes the perceived overall development relating to autonomy, including the maritime domain.

The bottom part of the roadmap lists identified key themes that are carried out over the course of the development. These include concern on the ethical issues (human factor, AI, et cetera), emphasis on the cyber security, development projects and IPR created in the companies (One Sea does not accumulate IPR) and ensuring the education is tailored to accommodate designing and operating with autonomous ships. The legislative work done nationally by the flag states and internationally is also seen as an ongoing activity.

In connection with the creation of the roadmap, One Sea has identified six main themes, which are the areas of focus for the ecosystem (Figure 3). The main topics are: technology, security (including safety), regulation, traffic control, ethics and operations. The themes are all partially overlapping. One Sea has also established working groups to ensure progress in some of the topics. Their purpose is to support fulfilling the 2025 vision of autonomous maritime ecosystem.

One Sea seeks to create industrial standards to efficiently develop the autonomous maritime transport. The standards may cover areas, such as safety, testing, data transfer, cyber security, data and systems architecture, artificial intelligence, et cetera. The objective is to harmonise the operational models within the industry with industry-specific regulations, where the rules and regulations drafted by the authorities may not correspond to the rapid development that the sector is experiencing. One Sea also collaborates globally with players within the field of autonomous maritime transport.

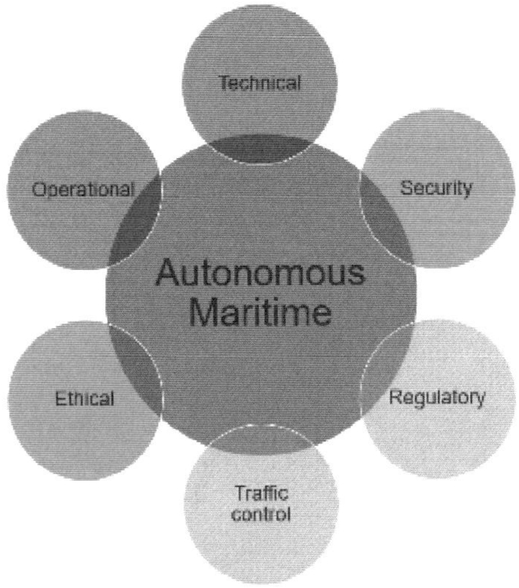

Figure 3: One Sea main themes.

3.2 THE STEPS TOWARDS COMMERCIAL AUTONOMOUS SHIPS

The following discusses the activities and the achievements of the One Sea ecosystem towards maritime autonomy.

3.2.1 Regulation

For international shipping, the IMO (International Maritime Organisation) is undertaking a scoping exercise of the regulatory implications of maritime autonomous surface ships (MASS) vessels [6]. Regulatory activities launched at the level of Rhine and/or Danube Conventions have also been started for inland waterway traffic.

One Sea is seeking to work with IMO via the flag states. One Sea worked with the Finnish flag state to produce two papers to MSC99 in May 2018. The first one is an info paper regarding the test area established in Finland and the other one is a working paper considering the definitions for levels and concepts of autonomy [7]. One Sea has been active in IMO MSC99, MSC100 and MSC101.

3.2.2 Testing and Validation

Thorough testing in authentic sea conditions is critical to ensure the functionality of systems and technology and to guarantee the required safety and reliability of the autonomous vessels of the future.

With the help of Finnish authorities, One Sea has established a test area in the coastal area of Finland in 2017. The test area titled, Jaakonmeri, was the first one in the world to be globally open to anyone wishing to test autonomous maritime traffic, vessels, or technologies related to it. The test area is managed and controlled by DIMECC Ltd. The test area was awarded by a Finnish Government Bureau, Centre for Economic Development, Transport and the Environment of Southwest Finland.

Jaakonmeri Test Area is located on the Finnish west coast outside the municipality of Eurajoki. The coordinates for the area are:

- NW 61 20.75 N 20 55.15 E
- NE 61 19.60 N 21 14.70 E
- SE 61 15.65 N 21 12.67 E
- SW 61 16.85 N 20 54.55 E

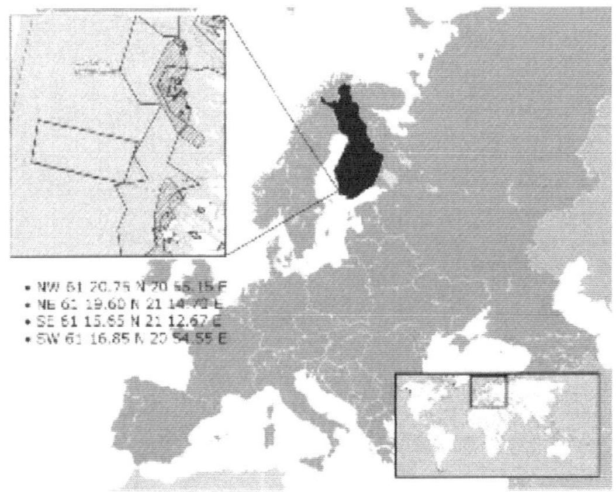

Figure 4: Jaakonmeri test area available for testing autonomous and unmanned surface vessels.

The longest side of the area to the north is approximately 17.85 km long and the western side is approximately 7.10 km long. The area is open water and offers opportunities to test also in ice conditions during the winter.
The area is closed for all maritime traffic except for the vessels that are tested and other vessel that are part of the test activities during the test period. The restrictions do not apply to the authorities and rescue operations. The test duration can range from one (1) to six (6) weeks. The sea trials will be adapted to other maritime traffic in the area, and it will be announced via the Vessel Traffic Service (VTS). During sea trials, the vessels will comply with the international rules of the waterways.

3.2.3 R&D Programs and Innovation

Research and development projects are an important part of the ecosystem. Although much of the technology required for realizing autonomous maritime is already available, there are items to be researched, clarified and optimized. One Sea ecosystem is a platform for R&D and innovation projects. These projects are joint efforts of One Sea companies, research organizations and companies outside One Sea ecosystem.
One Sea seeks to identify and engage with start-ups as well as agile companies (SME's) to identify innovation, foster growth and co-create technology, solutions and services (Figure 5).

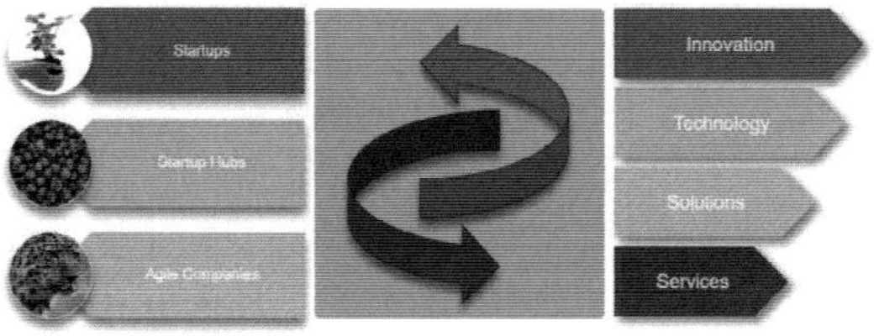

Figure 5: Collaboration with start-ups and SME's aims to find synergies and solutions.

3.2.4 Product and Service Creation

The most important activity in taking the autonomous maritime industry forward is the product and service creation by the industry. The ecosystem comprises both maritime and ICT companies, hence the development happens on multiple fronts. While companies have a long-term vision on where the maritime industry is headed, the services and products they are creating need to have a business case from the beginning. Thus, the companies are seeking to benefit from the low hanging fruits and build the portfolios incrementally. Among the first products to be introduced to the market are situational awareness systems with sensor fusion, automated docking systems and auto-crossing systems for short stretches. Situational awareness products observe the vessel's surroundings in ways beyond the capabilities of the human eye, using sensor technologies. Automated harbour operations are also available commercially. Figure 6 below highlights some of the recent milestones in the development of autonomous ships.

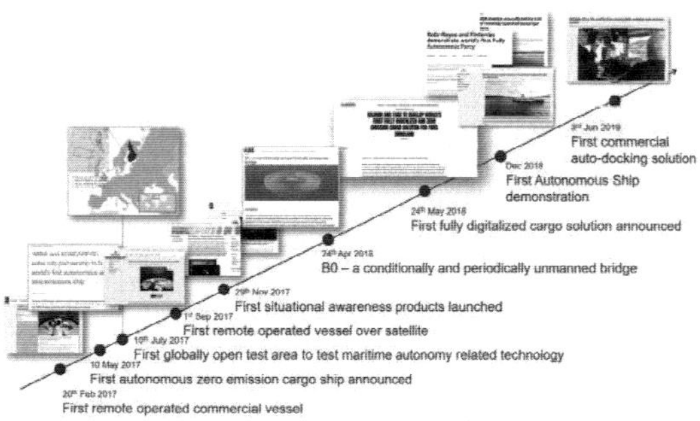

Figure 6: Some recent highlights in autonomous shipping.

While there have been several notable achievements during the past couple of years, four are mentioned here. The first one is Kongsberg Maritime's demonstration in February 2017, which involved remotely manoeuvring a 28m Svitzer Hermod tug form a remote operating centre (ROC), where the captain of the vessel was stationed. Although the tug was operated remotely, the vessel had fully qualified captain and crew on board throughout to ensure safe operation in case of system failure and to satisfy the current regulation. The technology involved in the demonstration included systems to give captain full awareness of surroundings in the ROC. Sensors included Radar, Lidar, camera and audio as well as DP system (Dynamic Positioning system). The demonstration was overseen by Lloyd's Register.

Second milestone to point out is the Wärtsilä's demonstration of remote piloting an offshore vessel few months later in September 2017 [8]. In this demonstration, an 80-metre platform supply vessel in the North Sea coast of Scotland. The remote navigation was done from Wärtsilä's office in San Diego, California some 8000 km away. The significance of the demonstration is in the communication between the vessel and the shore. The demonstration was carried out using standard bandwidth onboard satellite communication and no land-based technologies were. A DP system played also an important role in the successful 4-hour demonstration.

The third milestones to highlight involves the first product announcements around situational awareness. ABB announced its Ability Marine Pilot Vision product at the end of November 2017 [9]. Kongsberg Maritime published its Intelligent Awareness system in March 2018 [10]. Both systems improve the situational awareness thus improving navigational safety and enhance operational efficiency. These systems are support systems. Using them allows the crew to make more informed decisions. Both products are important steps towards autonomous maritime operations. Improving the maritime safety is one of the key features of the products.

Finally, the end of year 2018 marked an important milestone for development of autonomous transport system when ABB, Kongsberg and Wärtsilä demonstrated their capabilities in remote operations on a passenger ferry [11], worlds first fully autonomous ferry [12], and automated dock-to-dock solution [13].

One conclusion drawn from the presented milestones is that technology to enable autonomous operations is developing quickly. Focus on safety is crucial for the long-term business cases of maritime transport. While these milestones are not enough in fulfilling the promises of the future, the journey towards autonomous maritime transport is well on its way. This leads into One Sea's other important function: disseminating the message of maritime autonomy.

3.2.5 Disseminating the Message of Maritime Autonomy

One Sea strives to disseminate the message of maritime digitalisation autonomy and dispel concerns related to it. Much of this work is done by participating and speaking in conferences, seminars and discussing in different forums as well as publishing papers regarding maritime autonomy. It also includes meaningful collaborating with industry, authorities, class societies, research organizations and other stakeholders.

4. CONCLUSIONS

Transition from the conventional vessels will take time as the lifespan of the existing fleets are long. The initial implementations will be in local waters and short sea shipping. There will be vessel types where unmanned option may not suitable for a very long time. However, the safety increasing technologies will be highly applicable in all maritime operations.

5. REFERENCES

1. ROTHBLUM AM Human Error and Marine Safety. In U.S. Coast Guard Research & Development Center, Orlando, FL, 2000.

2. MSC VIII/11, 9.3.1964.

3. https://en.wikipedia.org/wiki/Self-steering_gear (accessed October b2018)

4. https://stories.kongsberg.com/content/story-behind-dynamic-positioning (accessed October 2018)

5. FINNISH MARINE INDUSTRIES Strategic Research Agenda for the Finnish maritime cluster 2017-2025. Ministry of Economic Affairs and Employment of Finland. 1.11.2016.

6. IMO http://www.imo.org/en/MediaCentre/MeetingSummaries/MSC/Pages/MSC-98th-session.aspx (accessed October 2018)

7. IMO MSC 99/5/6, 25.5.2018.

8. https://www.wartsila.com/media/news/01-09-2017-wartsila-successfully-tests-remote-control-ship-operating-capability (accessed January 2019)

9. http://www.abb.com/cawp/seitp202/cb99bdc1a0fe4de9c12581df002e46ca.aspx (accessed January 2019)

10. https://www.rolls-royce.com/media/press-releases/yr-2018/06-03-2018-rr-offers-ship-navigators-a-birds-eye-view-with-intelligent-awareness-game-changer.aspx (accessed January 2019)

11. https://new.abb.com/news/detail/11632/abb-enables-groundbreaking-trial-of-remotely-operated-passenger-ferry (accessed January 2019)

12. https://www.rolls-royce.com/media/press-releases/2018/03-12-2018-rr-and-finferries-demonstrate-worlds-first-fully-autonomous-ferry.aspx (accessed January 2019)

13. https://www.wartsila.com/media/news/28-11-2018-wartsila-achieves-notable-advances-in-automated-shipping-with-latest-successful-tests-2332144 (accessed January 2019)

6. AUTHORS' BIOGRAPHY

Päivi Haikkola is heading One Sea – Autonomous Maritime Ecosystem. One Sea is an alliance of (currently) nine companies and five organizations striving to create an autonomous maritime ecosystem by 2025. Päivi has a long history of working with marine related companies. She started her career in the nineties consulting and meeting hundreds of marine customers around the world. Since then she has worked with various strategic development projects in the pulp & paper and telecommunications industries apart from her many assignments in the marine industries. She has spoken at numerous conferences and seminars on various topics and has participated as a judge in international maritime awards panels, such as eg. Marine Propulsion Awards.
Päivi has held several administrative positions in the past and worked both at shipyards and suppliers to the marine industry. Among other positions, she has previously worked as Head of R&D, Marketing Director and Administrative Manager, she has also acted as a Board Member. Päivi received a Master's Degree in Economics (Business Administration) at Åbo Akademi University and a Master's degree in Naval Architecture at Helsinki University of Technology.

Jukka Merenluoto is Ecosystem Lead for One Sea, a high-profile business alliance leading the way towards an autonomous maritime ecosystem in the by 2025. He has lengthy background in ICT industry where he has held director positions and headed strategically important change initiatives in several companies in various industries including maritime, manufacturing, and telecom. He holds a master's degree in Computer Science. He is also MBA graduate from Henley Business School.

AERO-TRIANGULATION AND PHOTOGRAMMETRY IN SENSIBLE INFRASTRUCTURES – HOW THE INDUSTRY BENEFITS FROM LOW-COST 3D MODELLING CONDUCTED BY UNMANNED AERIAL VEHICLES (UAV)

M. Stein, Hapag Lloyd Centre of Shipping and Global Logistics, Germany

SUMMARY

Given recent technological advances, small size UAV provide a suitable mean for accurate 3D modelling of sensible (maritime) infrastructures. Such models provide a valuable source of information in general port safety and security planning as well as emergency response operations. This aspect is particularly crucial as ports belong to a nation's critical infrastructure, where its multiple supply chains and hinterland connections are sensitive to disruptions of any kind. Dealing with a rising number of disrupting externalities and increasing occupancy rates, ports have a constant need for affordable technological innovations with regard to risk management. Based on practical experiments, this paper reveals the possibility of generating realistic 3D models on the example of port facilities. It further reveals that initial states of a 3D model can be generated in a short timeframe, providing a valuable quick-response information source for first responder in emergency scenarios.

1. INTRODUCTION

With the majority to global goods being transported overseas, port facilities obtain a status of a nation's key infrastructure. Rising demands on efficiency of global transportation along with affordable transportation and handling prices results in "increasingly complex and economically expedient global supply chains for manufacturers and retailers" according to [12]. As a consequence logistic firms have become extremely vulnerable to the consequences of a disruption in the system [12] [15].

The 9/11 attacks raised the concern, that modes of transportation could be misused for terroristic purposes, leading to risk management of port operations becoming a central focus of maritime security programs both national and international [5] [23] The result was the ratification of the international ship and port facility security (ISPS) code in 2002, adopted in the European Union in 2004, to mitigate and control risks to maritime security. Studies, however, have shown that security, although uniformly being required, brings no added value to the port's customers and the willingness to pay surcharges for port security is extremely slight. With increasing complexity and competition on the one side and rising disruption risks on the other side, port operators see themselves in need for practical and affordable innovations of modern risk mitigation. Apart from the risk of a security incident, similar risk of disruption can be attributed to the aspect of safety. A breakdown in port operations after an external attack is likewise severe compared to a breakdown after a large accident e.g. a collapse of a container gantry crane. The findings of this contribution are not limited to maritime operations and provide a valuable basis for further sensible infrastructure protection research using low-cost UAV systems.

The use of unmanned vehicles to mitigate safety and security concerns in port operations was introduced by [29] [30] revealing a number of operational potentials. In line with this research, this paper introduces the concept of low-cost Unmanned Aerial Vehicles (UAV) based photogrammetry in ports for accurate realistic 3D modelling of maritime infrastructures. The research is structured as follows: First, a brief introduction of the concept of port security, photogrammetry and UAV operations is provided. Second, a decision frameworks based on an in-depth literature analysis as well as two field experiments is described. Third, the conducted experiments and data gathering processes as well as their results are discussed. At last, this paper's research and its contribution is concluded while pointing out future research opportunities.

1.1 PORT SAFETY AND SECURITY MANAGEMENT BACKGROUND

The ISPS-Code was the direct result of the 9/11 attacks, intending to globally harmonize maritime security. It was introduced in 2002 by the United States and ratified 2004 in the European Union with the directive VO (EG) 725/2004. The code is structured as an extension to the existing Safety of Life at Sea (SOLAS) framework initiated in 1974 as chapter XI-II. In practice, the intended uniformity of maritime security regulations among all kinds of port operations in a relatively thin framework led to confusion and a variety of best practice procedures among port operators given the absence of clear content within the code. It is argued that the post 9/11 urgency of reinforcing security regulations has led to a confused regulatory framework characterized by many initiatives not always coordinated [3].

The relatively openness of the ISPS regulations on the other hand allows for innovative procedures to be embedded and operated within port operations on a fast and relatively easy manner in consent with designated authority. This allows for example for unmanned vehicles to be easily embedded into everyday services assisting or even fully conducting ISPS operation in ports [29]. According to the ISPS-Code, each port facility must have a port facility security plan (PFSP) as basis for countermeasures mitigating maritime security risks. Those plans consist of printed structure plans of a port facility and allow only limited information in emergency situations.

1.2 ABOUT PHOTOGRAMMETRY

According to [4] available 3D information through photogrammetric techniques has long being utilized by geoscientists to visualize surfaces and extract topographic data from aerial photographs, that is also referred to as aerial-triangulation. Fundamentally, photogrammetry works on the basis that, two overlapping photographs allow to calculate the unique three-dimensional location of a set of given points shared in both photographs. New methods of photo-based 3D reconstruction reflect the evolution of photogrammetry from a highly specialized and expensive technique to a user-friendly and scalable methodology using commonly available devices [4]. The derivative 3D data can readily produce realistic 3D models of the area of interest while its models can widely be shared and easily be archived. In comparison to CAD models, 3D photogrammetric techniques provide a real life model of the actual status of the point of interest, therefore offering a great benefit for planning, analysing and management purposes. During the last decade, algorithmic advances in digital photogrammetry and computer vision have led to the technique of structure from motion (SfM) for image-based surface reconstruction, providing a great advantage in digital photogrammetry [11]. SfM enables automated model production from unconstrained imagery, for which metric accuracy is not the primary goal anymore [4]. This aspect is based on two conditions that SfM has overcome in the area of photogrammetric limitations. First, its advance of an increased tolerance to geographic changes in view points and second more robust matching algorithms that identify and reject errors when occurring. Advances in image matching and orientation, led to vast progress in dense matching results in digital elevation models (DEMs) with very high resolution [11]. Given the advances of SFM and DEMs, the requirements of current digital photogrammetry were reduced to an extent over the past years were current low-cost UAV already provide necessary means. Photogrammetric datasets derived from UAV platforms offer a cost-effective, ultra-high resolution alternative with a rapid acquisition time while delivering visual data in digital form [4].

1.3 DEVELOPMENTS OF UAV

Over the last decade unmanned aerial vehicles (UAV) have gained a significant role in conducting and assisting manned tasks of various backgrounds. Originated in military operations since WW2, UAV have entered civil environments and offer a vast potential for academic research and industrial application. With steadily increasing sensor power and increasing flight times while on the other side decreasing sizes and prices, UAVs gathered the attention of more and more small and medium sized companies. Between 2008 and 2018 the number of industrial produced UAV systems rose from 1.282 to 1.970 of which the number of civil and commercially used systems rose from 115 to 739 as displayed in table 1. Especially the development in the civil and dual purpose area underlines the industry's interest in unmanned systems from a market supply point of view. This paper differentiates UAV in general between multirotor systems, fixed-wing systems and vertical-take-off-and-landing (VTOL) systems. The focus of this paper lies on the application of low-cost multirotor systems of a widespread availability. For further classification details it is referred to [8].

Table 1: UAV Development 2008-2018

Application	2008	2009	2010	2011	2012	2013	2014	2015	2016	2017	2018
Civil/ Commercial	115	150	171	175	217	247	315	406	561	n.a	739
Military	578	683	631	674	548	564	609	600	644	n.a	543
Dual Purpose	242	260	283	318	353	392	447	502	559	n.a	539
Research	54	66	66	69	73	78	78	82	82	n.a	46
Developmental	293	329	301	310	187	172	162	140	161	n.a	103
Sum	**1.282**	**1.488**	**1.452**	**1.546**	**1.378**	**1.453**	**1.611**	**1.730**	**2.007**	**n.a.**	**1.970**

Source: Author referring to [25]

In the past few years, Unmanned Aerial Vehicles have been evolving as a cost-effective and flexible alternative to traditional mobile mapping platforms for providing high quality mapping products [10]. Modern platforms operate at low cost and enable flexible and frequent data-acquisition missions using a variety of sensors [7] [9] [11]. UAV photogrammetry opens various, new applications in the aerial domain and introduces also a low-cost alternatives to the classical manned aerial photogrammetry [7] [9]. It is argued that UAV operations for geo data acquisition are most suitable for small areas of interest where they bridge the gap between terrestrial and aerial data collection on a cost effective basis compared to traditional modes [21] as shown in figure1.

Figure 1: UAV operation matrix

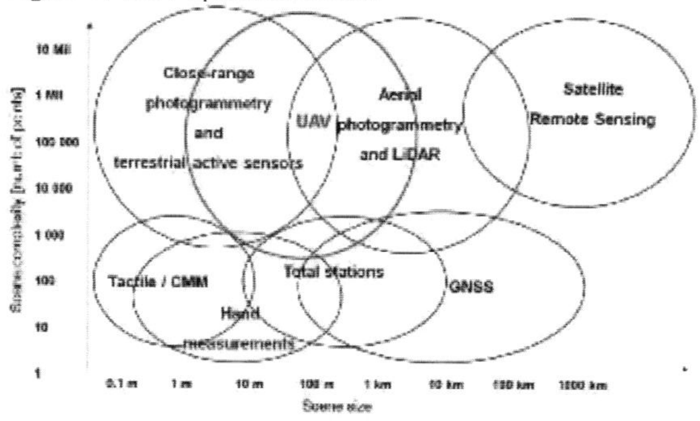

Source: [22]

UAV reveal a considerable range extension compared to close-range photography while offering a considerably cheaper solution compared to manned aerial photogrammetry. This paper argues that UAV photogrammetry opens a variety of practical applications in the industry and other civil services in the near future.

2. METHODOLOGY

The above stated technological advances in photogrammetric requirements being met by a growing number of UAV has opened the academic area on unmanned aerial 3D mapping over the last years. No contribution has so far introduced photogrammetry into industrial operations, let alone maritime logistics. It is this paper's intention to closing the gap by contributing to the existing literature in several ways. First a systematic literature research on UAV based photogrammetry is conducted. The generated insights provide the basis for a decision framework in the second step on operating photogrammetric attempts in the maritime domain. Lastly two field experiments are conducted generating 3D models of port facilities using a low-cost UAV system, those outcomes will be compared and discussed according to prior contributions.

2.1 DECISION FRAMEWORK

Conceptual frameworks as an initial source of academic contributions are widely accepted. They can be defined as a visual or written product that explaining the key factors, concepts, or variables to be studied and the presumed relationships among them [27]. Studies furthermore pointed out the importance of structured frameworks as a basis for future research for academic areas with limited existing literature [19] [27]. The possibility of conducting aerial photogrammetry using UAV instead of manned aircrafts inhibits both advantages and operational challenges. The latter are expressed by a stronger sensitivity of UAV specific characteristics as shown in figure 2.

Figure 2: UAV Photogrammetry planning characteristics

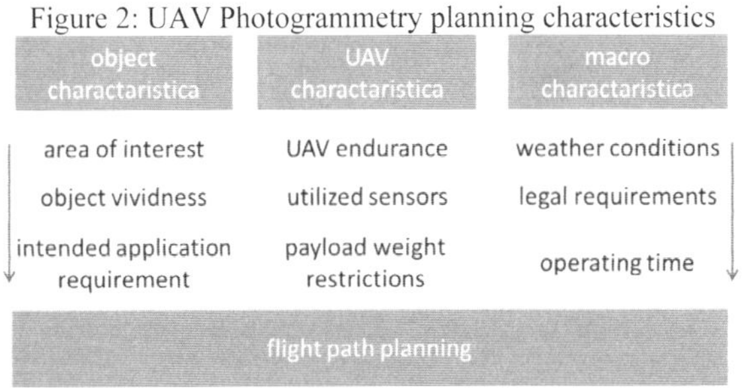

Source: Author referring to [2] [9] [18] [30]

The object characteristics define the basis for every photogrammetric operation and include the size of the area of interest, it's attributes as well as the elevation of specific points or buildings. The task lies in achieving optimal camera positioning relative to the object of interest, while operating with vegetation, topography, objective hazards, etc.. Combining UAV and SfM for geomorphologic applications, the actual surface representation has to be considered besides the systematic errors [11]. The target itself may appear to change between images due to wind shifting vegetation, or the movement of people and vehicles [4]. The higher the degree of movement of the object or around the object (vividness), the smaller the number of possible tie points within the generated images thus the more inaccurate the model. The intended application of the 3D model itself will also dictate the flight planning. A high quality model of a building for construction planning requires a different evaluation than a quick response low quality model for authority management (e.g. fire fighter or police).

The UAV characteristics of the specific aerial vehicle itself further define its operative capabilities. Especially when operating with low-cost systems, certain drawbacks or trade-offs need to be made [18]. The balance between the UAV payload capacity and the required accuracy for the final mapping product is one of the key challenges. For example, UAV payload restrictions might enforce the use of lower-grade (consumer-grade) sensors which can affect the quality of the model [6] and requires a careful mission planning for data acquisition [2]. The payload mainly affects the sensors being carried and the endurance of the system, meaning the time (the size of the area) that can be mapped with one battery load. Modern mapping software already allows for breaks within the mapping process to change batteries, but the danger of errors in the combination process of the software for each landing and starting is given, so that breaks during operations should be avoided. Further payload weight restrictions affect the carriage of for example safety gear such as anti-collision systems or parachutes that may be legally required in certain vivid areas.

On the macro characteristics, weather is the most obvious challenge for any low-cost and other UAV operation mainly defined by light, rain and wind. Direct sun exposure of the object creates a mixture of bright areas (overexposure) and shadows (underexposure) at the same time. It is therefore common to map a large object several times during the same day using the natural movement of the sun or, under ideal circumstances operate, with a light cloud cover (as done in this paper's experiments). Rain and wind generate a danger of increasing accuracy errors in the model or even losing the UAV system while operation. Similar studies operated their low-cost UAV between 3-5 m/s and experienced distinct differences to pre-set waypoint coordinates above 8m/s [21]. Prior contributions conducted UAV photogrammetry at wind gusts up to 10m/s without compromising the robustness of their model [4]. From experience 10m/s defines the border for any low-cost UAV operations in ports due to the risk of system loss and the loss of legal insurance coverage of the system and external fatalities above wind speed of 10m/s [30]. Rain during photogrammetric operations is to be avoided for it breaks the light between camera sensor and object and endangers the electricity of the relatively open UAV system. Legally, not every operation is allowed to be flown on an optimal path relative to the object or area of interest. Heterogeneity of UAV laws among countries and even states further differentiate a flight path planning from a macro characteristic. Time is another factor, affecting the planning and the quality of the model and is linked to the application requirement of the same. An emergency response low quality model has for the sake of this argument a lower time frame than a highly accurate marketing or management model.

2.2 FIELD EXPERIMENT

In order to analyse the use of a low-cost UAV system for aerial photogrammetry in port facilities, two experiments were conducted in July 2018. Experiment I was conducted on a cruise terminal in Lübeck, Germany while experiment II was conducted on a multipurpose berth facility in Rostock. Table 2 displays the experiment's background information.

Table 2: Experiment Environment

	Experiment I	Experiment II
Port type	Cruise terminal	Multi-purpose berth
Area size	Approx. 540 m²	
Ship in port	No	Yes
Vivid environment	Yes	No
Temperature	17,1 °C	18,5°C
Wind speed average	3,6 m/s	4,6 m/s
Wind speed std. dev.	0,58	0,41
Wind gusts average	5,3 m/s	7,4 m/s
Wind gusts std. dev.	1,12	0,69

Source: Author

No two ports are totally alike given their structural differences in general. Furthermore, when it comes to 3D mapping, which generally requires a motionless environment of the area of interest, additional barriers occur regarding to the port's position, everyday operation and other external factors. Experiment I was conducted with no cruise ship berthing at the port, resulting in the port not falling under ISPS regulations given its specific structure thus being open for public traffic creating a vivid environment consisting of anglers, stroller, private traffic etc.. The purpose was to test for UAV photogrammetry being capable of dealing with vivid environments when conducting 3D modelling. Experiment II was conducted under non-vivid conditions but with a ship berthing. The intention was to generate detailed information on board the ship and from the port itself at once. The ship was a chemical tanker generally known for their complicated on-board structure of pipes and valves. Water is known for being a source of disturbance for tie point generation within 3D modelling, so the intention was to test for the model accuracy around the ship hull near the surface.

3. DATA GATHERING

Collecting visual data from unmanned systems for accurate 3D modelling requires certain principles with regard to flight path and picture quality. Sources, however, vary with regard to the required overlap rate of the pictures taken from 50-60%[1] to 60-80% [13]. The accuracy of a photo-based model depends upon the scale and resolution of the input images, the precision and distribution of matched image points, and how much they overlap [4]. To enhance the data quality, effects that reduce textural variability within images should be minimized during acquisition. Increasing angles of convergence between overlapping images will tend to increase reconstruction accuracy up to a point, but will eventually prevent matching due to the surface texture appearing too dissimilar in images from different directions [4]. Within traditional airborne image flights, pictures are taken in nadir direction towards ground and in parallel flight strips [13]. Studies revealed, that automated image matching does not perform well with angular changes > 25-30° between images [20] and should be limited to 10-20° for overlapping photos [4]. Additional common issues preventing algorithms from resolving coincident points include homogeneous surface texture, changes in the target, and changes in illumination [4]. On average 20-40% of the pictures taken have to be eliminated in practice because they are not matching the software algorithm requirements [21].

3.1 UAV USED

UAV utilized for mapping applications are usually equipped with direct geo-referencing units, passive and/or active optical imaging systems as well as a lightweight active optical imaging unit and one or more passive optical imaging systems [18]. The UAV used in this contribution fulfils the above defined requirements with specifications stated in table 3.

Table 3: UAV Specifications

weight	1.388 g
Size	350 x 350 mm
Acceleration speed	5-6 m/s
Flight speed	50-72 km/h
Max wind resistance	10 m/s
Max flight time	30 min
Optical obstacle avoidance area	0,2 – 7m
Operative frequencies	2,400 - 2,483 GHz and 5,725 - 5,825 GHz
Max transmitting range	7 km

Source: Author referring to UAV manufacturers website

The UAV used to conduct the experiments was a low-cost system at costs below 7.500 € which undercuts system prices of prior contributions conducting UAV driven photogrammetry. [21] conducted the first photogrammetry approach using a comparable prized low cost UAV. They argued that a considerable advantage of modern UAV systems lie in the inclusion of geo referencing technology on-board the system. According to them this aspect results in no signalization, no local survey of reference points in the field as well as no determination within the point cloud is needed. Using a direct geo-referencing unit on board the UAV is employed for navigation of the drone and its position via optical data transmission using GPS data generated during the flight.

3.2 CAMERA USED

The capability of a conventional camera from a low-cost UAV system offers a potential suitable for aerial photogrammetry when combined with modern software and calculation algorithms. The sensor specifications of the camera used are provided in table 4 below.

Table 4: UAV Specifications

Sensor	1" CMOS
Quality	20 MP
Objective	Field of view 84°, 8.8mm/24mm
Shutter speed mechanic	8 – 1/ 2.000s
Shutter speed electric	8 – 1/ 8.000s
ISO	100-3.200 (auto); 100-6.400 (manual)
Focal aperture	0,7

Source: Author referring to UAV manufacturers website

The camera used had a fixed lens without zoom function. Contributions have shown, that movements of a zoom-lens can result in changes in the camera geometry because of flexible lens alignment [4] [17] [32]. Furthermore, uncertain camera calibration is determined to be a possible cause for insufficient correction of image distortion [11]. Prior contributions stated that the use of small or medium format cameras enforce the acquisition of a higher number of images in order to achieve the same image coverage at a comparable resolution compared to larger camera systems [24]. Although agreeing on that aspect from a camera driven aspect, the technological advance in software algorithms compensate this aspect to an extent, that even small camera systems deliver a considerable useful result.

3.3 SOFTWARE USED

3D software experienced tremendous advances during the last decade of aerial photogrammetric. Algorithms that combine local and global characteristics have been developed, such as semi-global matching (SGM), which performs pixel-wise matching minimizing matching time and efforts [16]. Basic SfM tools calculate relative image positions and camera orientation fully automatically providing a huge advantage of reduced mapping efforts allowing models to be effectively constructed from a wide variety of imagery [24] [28]. Contributions already evaluated the different software performances for UAV based 3D mapping [11] [31]. It is shown, that more complex software (such as pix4D) allows the input of the camera's position and orientation in the bundle block adjustment process, thus enhancing the model quality. Among the most salient reasons for conducting this paper's research with Pix4D was the detailed report functions of the program including scores, datasets, matching and geo-referencing quality, and detailed quantitative measurements of the AAT, BBA or ground control points accuracy [31]. Furthermore, all photogrammetric calculations and measurements are fully automated within the software allowing for a fast application in emergency situations. Pix4D is a commercial none open source software with a number of different features and advanced algorithms for unmanned photogrammetry combined to an easy to use interface. The image processing of this paper was performed with Pix4D Mapper 4.3.15

3.3 DATA ACQUISITION

The data acquisition of this paper's experiments follows the 9-step schematic provided by [4] as displayed in figure 3. The preparation of the workflow i.e. the consideration of the UAV system, the camera used and the software applied were already introduced in chapter 3. The fieldwork requires the establishment of ground control points across the area of interest. Ground control points help defining the area where no significant characteristics can be distinguished such as soil or landscape mapping. In this paper's experiments no additional control points were needed as buildings provided enough area for tie points. The first stage of a SfM-based 3D reconstruction involves the analysis of the individual images for distinct image features that can be matched to their corresponding features in other images within the collection [4]. This step is often referred to as "bundle adjustment" that aligns the images using an arbitrary coordinate system. One of the major advances in SfM is that conventional coordinate systems have turned into large networks of matched points. These networks consist of relative locations of each camera, the camera parameters for each image, the collection position and orientation for each input image, and the 3D coordinates for each matched feature [4]. This process optimizes all variables to produce a consistent 3D model, with minimized overall residual error, the sparse point cloud. A point cloud as preliminary outcome is the most suitable product that could simultaneously consider the synergistic characteristics of the involved passive and active imaging systems [18]. Several studies already pointed out the high potential of the combination of UAV image flights and the SFM processing chain, particularly with regard to urban objects and terrain modelling [13] [21] [24]. A transformation is subsequently applied to the sparse point cloud, resulting in thousands of real-world coordinates linked to points in each image using a Delaunay triangulation. Automated dense matching techniques that have been introduced in various contributions [14] [16] [26] are then utilized. The dense matching procedures benefit from the direct geo-referencing capability and the system calibration parameters of mated points in overlapping imagery resulting in a more dense point cloud [4]. Direct geo referencing offers the advantage that no signalization, no local survey of reference points and no determination within the point cloud is needed. These geo-references are usually provided by either real-time kinematic (RTK) global navigation satellite system (GNSS) points or LIDAR.Although the use of RTK techniques would improve the quality of positioning, the system would become too complex, expensive and heavy to range within the boundaries of a low-cost UAV system [24].

Lidar on the other hand is not able to provide a photorealistic model due to missing data. This paper applies GNSS data for geo-referencing during the flight. Although inferior in RTK or LIDAR quality, GNSS observations are seen suitable for fast data acquisition and mapping during emergency response [33].

Figure 3 –UAV Photogrammetry workflow

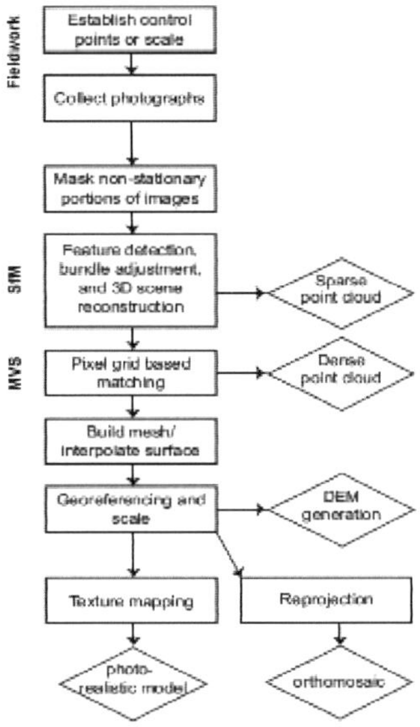

Source: [4]

Finally, a digital elevation model (DEM), ortho-rectified mosaics and photo realistic models can be derived using geo-referenced aerial data. 3D models at that stage can furthermore be textured and exported into common 3D visualization formats (e.g..obj,.ply,.pdf, etc.) allowing a fast distribution of the generated knowledge [4].

3.4 DATA DESCRIPTION

As displayed in Annex I the data acquisition conducted in 35 minutes (experiment I) and 31 minutes (experiment II) in total. Different flight path were used to cover the specific area and the structures of interest resulting in five paths for experiment I (with a more complicated area including trees and civilian houses) and three paths for experiment II (with easy and uniform building structures). Both experiments were conducted in absence of direct sun exposure to avoid shadow creation within the area of interest that could lead to increasing matching errors among the data as discussed in chapter 2.

4. RESULTS AND DISCUSSION

Detailed information of the experiments is provided in annex I and II. Following the modelling path of [4], the upper left picture in figure 4 shows the collection of photographs on the object of interest. 831 pictures in total were gathered within a flight time of 35 minutes of which 781 were calibrated by the software and integrated into the final model, resulting in a 93% matching rate. The average ground sampling distance is 0,91cm and the relative difference between initial and optimized internal camera parameters is 0,51%. The upper right picture of figure 4 reveals the sparse

Figure 4 Experiment I: 3D modelling process

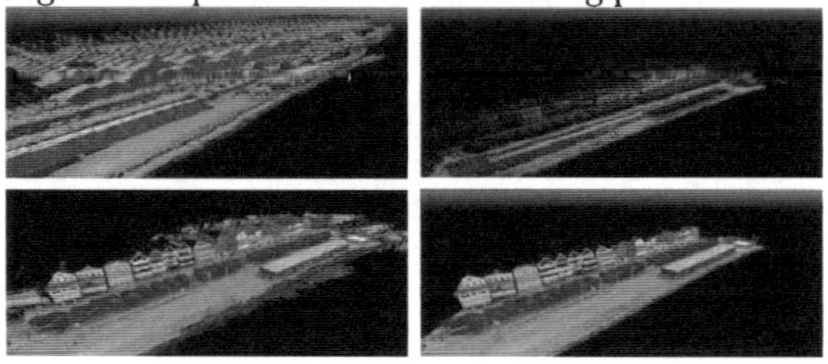

Source: Author (2019)

point cloud generated from the data consisting of a median of 39.472 key points per image in total and 7.103 matches per calibrated image. The point cloud took 80 minutes to be calculated. The BBA, displayed in the lower left picture of figure 4, took 486 minutes to be processed including a total of 6.121.848 2D key points and 1.731.225 3D key points. The time for textured mesh generation was 21 minutes. The final model is shown in the lower right picture consisting of 4 tiles, a number of 41.696.519 densified points and an average density per m^3 of 6.187 points. Minor manual corrections were conducted for displaying purposes, such as erasing of noise.

Using the same technique as experiment I, 417 pictures in total were gathered within a flight time of 32 minutes and were fully integrated into the final model, resulting in a 100% matching rate. The average ground sampling distance is 1,06cm and the relative difference between initial and optimized internal camera parameters is 0,57%. The sparse point cloud generated from the data consisting of a median of 38.740 key points per image in total and 13.611 matches per calibrated image. The point cloud took 65 minutes to be calculated. The BBA was conducted in 500 minutes to be processed including a total of 5.814.273 2D key points and 1.819.000 3D key points. The time for textured mesh generation was 15 minutes. The final model consisting of 1 tile, a number of 16.782.877 densified points and an average density per m^3 of 3.990 points. Minor manual corrections were conducted for displaying purposes, such as erasing of noise.

Figure 5 Experiment I: 3D modelling process

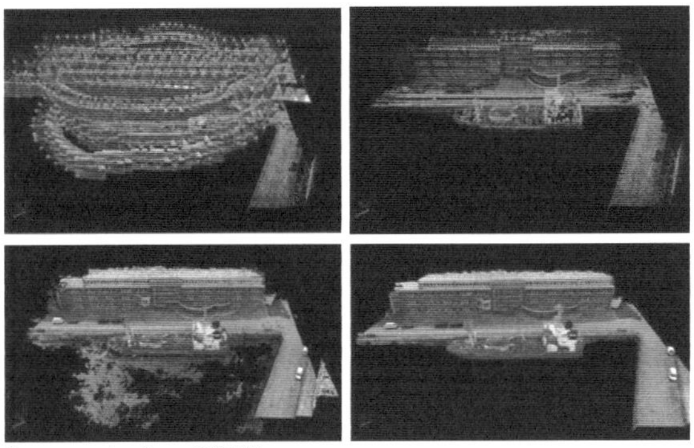

Source: Author (2019)

The final model of experiment II revealed errors in identifying the ship's on board structure above the surface of the water. Additional test using additional software provided more promising results as displayed in figure 6. The software added was Context Capture version 4.4.7. The trial version's limitations of the programme did not allow for a full test of the above stated experiments in similar wealth of details, which would furthermore have exceeded this paper's research boundaries. It is shown, however, that improvements of the outcome are feasible given modern software solutions and that additional research in this area is necessary.

Figure 6 Experiment I: 3D modelling process

Source: Author

4.3 DISCUSSION

This paper successfully reveals the ability to construct realistic 3D models of port infrastructures using low-cost UAVs. Two different scenarios were tested in line with prior contributions of comparable photogrammetric approaches. In line with [4] [21] wind speed of less than 10m/s occurred during the experiment. Wind gust of up to 9,8 m/s were measured but did not significantly interfere with the model quality and the matching results. These results show that when using larger UAV models, compared to this paper's UAV, good results could be achieved above the 10 m/s wind speed threshold. Studies argue that on average 20% to 40% of the pictures taken have to be eliminated in practice for not matching the software algorithm requirements [21] and. This paper's results do not go in line with this estimation when reaching 93% and 100% accuracy. This result further indicates the high advance of modern 3D software capabilities and the high resolution of low-cost UAV camera systems. Following [18] having low-cost imaging and geo referencing sensors on board UAVs with limited payload capability, the necessity of efficient data processing techniques and the diversity of potential users with insufficient mapping-related expertise needs to be addressed.

As already pointed out, quality of the model and time needed to generate the model are two fundamental aspects for the actual use of the photogrammetric data. This paper focussed on the fact of using low-cost UAVs for quick response operations as well as detailed models for general management in a later processing step. The experiments revealed very detailed point clouds of the two areas in were generated after about half an hour flight time and 65-80 minutes of processing. This paper argues that additional research on this matter will further reduce the processing time, and so providing a valuable quick response mechanism of aerial based infrastructure imagery.

5. CONCLUSIONS

This paper provides a valuable contribution to the UAV photogrammetry and maritime security research by revealing practical means to conduct aerial based 3D mapping of port facilities. It is the first time, that UAV based photogrammetry is tested in sensible infrastructure using easy to use and widely available low cost UAV systems. In line with [24] the great advantage of UAV systems is the ability to quickly deliver high resolution image information and to allow a rapid response in a number of critical situations where immediate access to 3D geo-information is crucial. Given the sensible infrastructure of port facilities and the high sensitivity to disruption within ports, low-cost UAV systems prove to be a valuable contribution to safety and security risk management.

6. REFERENCES

1. ABDULLAH, Q., BETHEL, J., HUSSAIN, M., MUNJY, R., 2013, Photogrammetric project and mission planning. In: MCGLONE, J.C. (Ed.), Manual of Photogrammetry. American Society for Photogrammetry and Remote Sensing, 1187-1220.

2. ABER, J.S., MARZOLFF, I., RIES, J., 2010, Small-Format Aerial Photography: Principles, techniques and geoscience applications, 1 edition. Elsevier Science, Amsterdam.

3. ACCIARO, M and SERRA, P., 2013, Maritime supply chain security. a critical review. IFSPA

2013, Trade Supply Chain Activities and Transport: Contemporary Logistics and Maritime Issues, 636-651.

4. BEMIS, S.P., MICKLETHWAITE, S., TURNER, D., JAMES, M.R., AKCIZ ,S., THIELE, S.T., BANGASH, H.A., 2014, Ground-based and UAV-Based photogrammetry: A multi-scale, high resolution mapping tool for structural geology and paleoseismology. Journal of Structural Geology 69, 163-178.

5. BICHOU, K., 2008, Security and risk-based models in shipping and ports - review and critical analysis. International Transport Forum Discussion Paper 2008-20.

6. CHIANG, K.W., TSAI, M.L., CHU,C.H., 2012, The development of an UAV borne direct georeferenced photogrammetric platform for ground control point free applications. Sensors 12 (7), 9161–9180.

7. COLOMINA, I. and MOLINA, P., 2014. Unmanned aerial systems for photogrammetry and remote sensing: a review. ISPRS Journal of Photogrammetry and Remote Sensing, 92, 79–97.

8. DALAMAGKIDIS, K., 2015, Classification of UAV. In *Handbook of Unmanned Aerial Vehicles*,edited by Valavandis K.P. and Vachtsevanos G.J., 1st ed. (Dordrecht: Springer, 2015).

9. EISENBEISS, H., 2009, UAV photogrammetry. Doctoral dissertation ETH Zurich, Zurich, Switzerland.

10. EISENBEISS, H., 2011, The potential of Unmanned Aerial Vehicles for mapping. Proccedings of the Photogrammetric Week 2011, Stuttgart, Germany, 135–145.

11. ELTNER, A. and SCHNEIDER, D., 2015, Analysis Of Different Methods For 3d Reconstruction Of Natural Surfaces From Parallel-Axes UAV Images. The Photogrammetric Record 30(151), 279–299.

12. FLYNN, S. E., 2006, Port security is still a house of cards. Far Eastern Economic Rev. January/February

13. FRITZ, A.; KATTENBORN, T. and KOCH, B., 2013, UAV-Based Photogrammetric Point Clouds - Tree Stem Mapping In Open Stands In Comparison To Terrestrial Laser Scanner Point Clouds. International Archives of the Photogrammetry, Remote Sensing and Spatial Information Sciences (40), 141-146.

14. GEHRKE, S., DOWNEY, M., UEBBING, R., WELTER, J., LAROCQUE, W., 2012. A multi-sensor approach to semi-global matching. ISPRS - International Archives of the Photogrammetry, Remote Sensing and Spatial Information Sciences 39 (B3), 17–22.

15. HECKER, J.Z., 2002, Port security: nation faces formidable challenges in making new initiatives successful. U.S. General Accounting Office.

16. HIRSCHMÜLLER, H., 2008, Stereo processing by Semi-Global Matching and Mutual Information. IEEE Transactions on Pattern Analysis and Machine Intelligence (PAMI), 30(2), 328–341.

17. JAMES, M. R. and ROBSON, S., 2014, Mitigating systematic error in topographic models derived from UAV and ground-based image networks. Earth Surface Processes and Landforms, 39(10), 1413–1420.

18. LARI, Z. and EL-SHEIMY, N., 2015, System Considerations And Challendes In 3d Mapping And Modeling Using Low-Cost UAV Systems. The International Archives of the Photogrammetry, Remote Sensing and Spatial Information Sciences 40 (3), 343-348.

19. MILES, M.B. and HUBERMAN, A.M., 1994, Qualitative Data Analysis: A Sourcebook of New Methods. Beverly Hills, CA: Sage.

20. MOREELS, P., PERONA, P., 2007, Evaluation of features detectors and descriptors based on 3D objects. International Journal of Computer Vision (73), 263-284.

21. NEITZEL, F. and KLOSNOWSKI, J., 2011, Mobile 3d Mapping With A Low-Cost UAV System. International Archives of the Photogrammetry, Remote Sensing and Spatial Information Sciences 38, 1-6.

22. NEX, F. and REMONDINO, F., 2014, UAV for 3D mapping applications: a review. Appl. Geom. 6 (1), 1–15.

23. PALLIS, A.A.and VAGGELAS, G.K., 2007, Enhancing port security via the enactment of EU policies. In: BICHOU,K. , BELL, M. G. H. and EVANS, A., Risk management in port operations, logistics and supply chain security ,303-334, London: Informa.

24. REMONDINO, F., BARAZZETTI, L., NEX, F., SCAIONI, M.AND SARAZZI, D., 2011, UAV Photogrammetry For Mapping And 3d Modeling – Current Status And Future Perspectives. International Archives of the Photogrammetry, Remote Sensing and Spatial Information Sciences 38, 25-31.

25. RPAS Yearbook (2018), RPAS: The Global Perspective 15th Annual Edition – 2018. UVS International.

26. SEITZ, S., CURLESS, B., DIEBEL, J., SCHARSTEIN, D., SZELISKI, R., 2006, A comparison and evaluation of multi-view stereo reconstruction algorithms. Proc. IEEE Conf. CVPR'06 (1), 519-528.

27. SHIELDS, P. and RANGARAJAN, N., 2013, A Playbook for Research Methods: Integrating Conceptual Frameworks and Project Management. Stillwater, OK: New Forums Press.

28. SNAVELY, N., SEITZ, S.M., SZELISKI, R., 2007, Modeling the World from Internet Photo Collections. International Journal of Computer Vision 80 (2), 189-210.

29. STEIN, M., 2018a, Integrating Unmanned Vehicles in Port Security Operations: An Introductory Analysis and First Applicable Frameworks. Oceans Yearbook 32, 556–583.

30. STEIN, M., 2018b, Conducting Safety Inspections of Container Gantry Cranes Using Unmanned Aerial Vehicles. Dynamics in Logistics LDIC 2018, 154-161.

31. VISOCKIENE, J.S.; BRUCAS, D. and RAGAUSKAS, U., 2014, Comparison of UAV images processing software. Journal of Measurements in Engineering, Vol. 2 (2), 111-121.

32. WACKROW, R. and CHANDLER, J. H., 2011, Minimising systematic error surfaces in digital elevation models using oblique convergent imagery. Photogrammetric Record, 26(133), 16–31.

33. ZHOU, G., 2009, Near real-time orthorectification and mosaic of small UAV video flow for time-critical event response. IEEE Trans. Geoscience and Remote Sensing, 47(3), 739-74.

7. AUTHORS BIOGRAPHY

Michael Stein holds the current position of researcher at the Hapag Lloyd Centre of Shipping and Global Logistics (CSGL). His previous experience includes a vocational education as shipping agent as well as a M.Sc. degree in Management with focus on maritime logistics. His practical work experiences range from ship owner- over classification society- to consultancy companies. Furthermore, he is a licenced UAV pilot, a trained Port Facility Security officer (PFSO) and an independent consultant.

Annex I

	Flight details			Wind details			Weather Details					
Flight/Service Date	Date	Air Time	Max Altitude (Meters)	Max Speed (m/s)	Avg Wind (m/s)	Max Gust (m/s)	Ground Weather Summary	Ground Tem-perature (c)	Cloud Cover	Humidity	Rain Rate mm/h	Rain Chance
	Experiment II											
05.07.2018 08:15	05.07.2018	00:01:44	42,4	1,0	2,9	3,60	Mostly Cloudy	16,70	75%	82%	0,0	0%
05.07.2018 08:37	05.07.2018	00:06:45	43,6	8,8	4,1	5,50	Mostly Cloudy	16,90	74%	82%	0,0	0%
05.07.2018 08:47	05.07.2018	00:11:48	43,6	7,6	2,9	4,40	Mostly Cloudy	16,90	74%	82%	0,0	0%
05.07.2018 09:05	05.07.2018	00:09:17	36,7	6,1	3,9	6,30	Mostly Cloudy	17,10	74%	81%	0,0	0%
05.07.2018 09:40	05.07.2018	00:05:17	43,7	5,1	4,2	6,80	Mostly Cloudy	17,70	73%	78%	0,0	0%
Average			**42,0**	**5,72**	**3,6**	**5,32**		**17,06**	**74%**	**81%**	-	-
Std. dev.			**2,69**	**2,68**	**0,58**	**1,18**		**0,34**	**0,01**	**0,02**	-	-
	Experiment II											
18.07.2018 07:06	17.07.2018	00:01:13	34,8	3,2	4,7	6,4	overcast	18,5	99%	89%	0,0	0%
18.07.2018 07:18	17.07.2018	00:11:53	35,6	9,8	5,1	7,9	overcast	18,5	99%	89%	0,0	0%
18.07.2018 07:31	17.07.2018	00:18:16	25,7	8	4,1	7,8	overcast	18,4	99%	89%	0,0	0%
Average			**32,0**	**7**	**4,6**	**7,4**		**18,5**	**99%**	**89%**	-	-
Std. dev.			**4,49**	**2,79**	**0,41**	**0,68**		**0,05**	**0,00**	**0,00**	-	-

Source: Author

Annex II

	pictures taken	Pictures processed	processing ratio in %	flight time of UAV in minutes	average ground sampling points in cm	delta initial and optimized internal camera parameters in %	processing time sparse point cloud in minutes	keypoints of sparse point cloud (median)
Experiment I	831	781	93	35	0,91	0,51	80	39.472
Experiment I	417	417	100	32	1,06	0,57	65	38.740

	matches per calibrated image	BBJ processing time in min.	BBJ number of 2D Keypoints	BBJ number of 3D Keypoints	textured mesh proces.time in min.	densified points	average density in points per m^3	
Experiment I	7.103	486	6.121.848	1.731.225	21	41.696.519	6.187	
Experiment I	13.611	500	5.814.273	1.819.000	15	16.782.877	3.990	

Source: Author

THE CHALLENGES AND OPPORTUNITIES FOR DEVELOPING DIGITAL TWINS TO SUPPORT MARITIME OPERATIONS

S. Willmore, G. Mercer, J. Rigby, BMT, UK

SUMMARY

The concept of Digital Twins provides a conceptual way of improving the through life support of Marine platforms. The collation of machinery information through connected smart sensors and connectivity ashore enables advanced data analytics providing insights into the causes and trends for platform or fleet performance.

The opportunities are vast however, the challenges are also considerable. Increased sensor connectivity similar to the Internet of Things can lead to considerable cyber security concerns, and the amount of information available will need to be presented to the user in a suitable but succinct manner possibly using visualisation technology. How this vision is brought together alongside legacy processes, tools and data centres is non-trivial but modern solutions and approaches make a Digital Twin possible.

1. INTRODUCTION

The concept of Digital Twins (DTs) provides a conceptual way of improving the through life support of Marine platforms. The collation of machinery information through connected smart sensors and connectivity ashore enables advanced data analytics providing insights into the causes and trends for platform or fleet performance. This paper will explore the opportunities and the challenges to creating a DT in maritime shipping.

2. DIGITAL TWINS

2.1 WHAT WE MEAN BY DIGITAL TWIN

A DT can be thought of as a virtual representation of a product, an asset or a system. Whilst this is usually taken to mean physical objects, we also consider the DT scope to include systems of systems, processes and relationships. This wider definition enables enterprise level opportunities for DT to transform, optimise or disrupt the business.

A Digital Twin usually comprises of a collection of various computer software based models, together with information and processes pertinent to management of the object in question.

The business objective for creating a DT is to improve through-life management of the represented object, using the gathering information to provide insights into how to improve the operation or support of the object.

As well as the objects current build state, the DT can also feature dynamic data, taken as measurements from the (remote) operational object. The data can be processed and then formed into various DT outputs such as models, dynamic or discrete simulations, analytical models, sensor data models as well as graphic and report outputs.

In all cases it is the application of a large dataset with other sources of information that gives the DT concept its value for organisations.

In summary, the typical features of a DT are:

- Product based
- An output of the business
- Contains documentation and models
- A Product Configuration record
- Enables Change Impact and recording
- A digital product that resides on hardware and accessed via software
- Can be supplemented with live measurements from the operational object

2.2 WHY DIGITAL TWINS MATTER IN MARINE INDUSTRY 4.0

The increase in available computing power, improving connectivity at sea and the constant drive to manage costs is enabling new business functions whilst transforming existing ones.

With greater access to operational information harvested from ships, ports, the oceans etc. ship operators can now gain actionable insights not previously possible. By combining datasets such as equipment running times, vessel locations, meteorological and oceanographic details it is now possible to refine the maintenance of systems between individual ships in a fleet based on their unique operating histories.

Digital Twins are the vehicle for this business improvement. By using and combining all the available data the application of data analytics, DTs can provide greater insights and value to business reducing unnecessary activities, minimising unexpected events and saving costs.

For the customer, greater insights into their operations within the larger systems of systems business environment will enable them to make more informed management decisions. Specifically for Ship Operators the establishment of Digital Twins can allow for the analysis needed about a ship or fleet to be done away from the platform by specialists, which reduces the need for training and expertise to be deployed at sea.

Business stakeholders usually favour the efficiency of online experiences over more direct face-to-face methods. An efficient online DT platform can align all aspects of client activity for through-life support of a maritime product. This helps to reduce the service time, costs, and improves the likelihood of client repeat business in the future.

Figure 1 illustrates the formation and application of the DT concept to a maritime product.

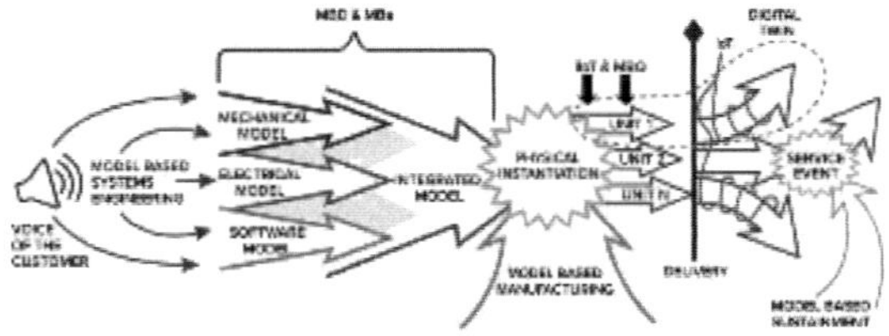

Figure 1: DT Formation and Service Use Using Product Data [8].

2.3 DIGITAL TWINS USE IN MARINE INDUSTRY 4.0

When introduced, Digital Twins are a type of enterprise digital transformation that can be applied to any part of a maritime organisation whether this be front office, back office or in the wider supply chain, Figure 2.

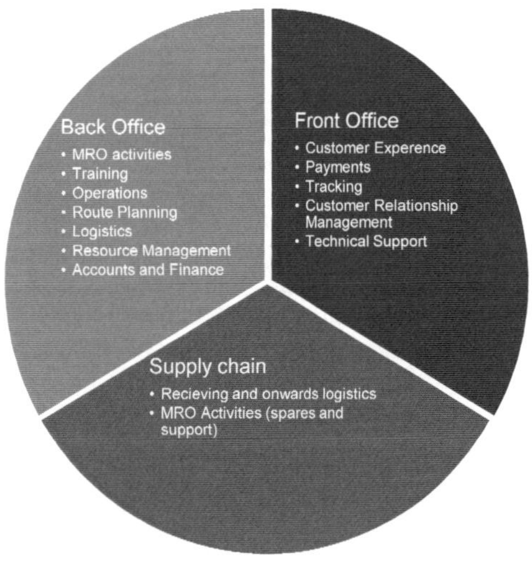

Figure 2: Examples of Where Digital Twins Could be Used to Support Commercial Marine Operations.

2.4 AVAILABILITY OF DIGITAL TWINS

Much of the underlying technology needed to create Digital Twins is readily available with reducing costs as wider adoption is taken up. In sectors where Digital Twin related technology has been adopted typical savings of around 15-25% have been realised [9].

DT adoption is dependent upon the business recognising a need to change, and energy to undertake a transformation programme.

With wider adoption the need to transform will become pressing for all organisations, in order to maintain parity with competitors by providing an operational advantage to deliver competing returns for investors.

With DT providing wide practical applications across the product life-cycle, e.g. testing prior to manufacture, improving operating efficiency and productivity, managing assets in real time and understanding data to provide service improvements, DT adoption in any one organisation is likely to be staggered across product lines, focusing on where the best business returns lie.

3. OPPORTUNITIES FOR DIGITAL TWINS TO REDUCE COSTS

The cost of managing shipping operations is more complex and costly than ever. By maximising the use of digital technologies and improved connectivity these costs can be reduced and the impact and likelihood of risks occurring can be minimised.

3.1 CREW

Whilst a significant cost to the operation of ships, the quality of crews have a significant long term impact on the cost of the platforms, as well as risk management issues. It is common that most maritime failures are caused by some form of human intervention [1].

Crew training is an area of concern and during difficult economic conditions recently this has suffered. Whilst mandatory training is being delivered there is a feeling that companies are being left to provide additional support in their own training centres [1].

Digital Twins of platforms can allow for more realistic hydrodynamic ship models to be used to allow immersive or standalone training models. By making these models and simulators more accessible crew training can be enabled not just in training centres but also on board and ashore.

Greater digitisation of ship systems and availability of information on the layout of ships also provides training opportunities. By making use of synthetic training platforms, either as VR or on conventional screens, engineers can learn about their ships ashore or on-board where a real world walk-through of procedures is impractical.

A further challenge to seafarers will be the pace of change on platforms. Greater connectivity will bring new challenges such as the need to maintain system security measures but this also allows for systems to be updated whilst still at sea. Moving to a more flexible training model where applications are designed to be easy to train on, will make the need for on-board training solutions even greater.

Digital Twins as a part of training will help promote better crew wellness as individuals will be able to train for longer and with greater repetition if they desire which helps to reduce stress and reduce the occurrence of mental health disorders and accidents.

3.2 CREW TRAVEL

As part of greater connectivity and resource management, the ability to help seafarers travel more effectively is possible. Door-to-door management of personnel making long journeys from ships to their homes is now possible which reduces stress on the individuals.

Sharing information about ports or journeys can also allow travel teams to avoid stressful routes or to promote other ones.

Together through making use of connectivity and sharing knowledge better the travel of crew members can be made less tiring, less stressful which helps to ensure better operational safety on arrival and improved health and well-being for the individuals.

3.3 FUELLING AND LUBRICATION

The use of fuels and lubricants is a major cost driver for ship operators. With environmental pressure to reduce emissions, changes have already been made to how ships operate and which fuels are chosen [10].

As ship operators move to low-sulphur fuels it is expected that problems with quality and compatibility of fuels may be an issue. Also, the impact on individual engine systems may increase failure rates. A Digital Twin able to track fuel usage from source as well monitor engine performance over time to spot trends will be important to identify and understand the long term impact of decisions being made [1].

The type or formulation of fuel is not the only factor as the efficiency of the ship to use its fuel also has impact on the total cost of a voyage. Whilst vessel efficiency can be improved by changing engines, components and operating procedures understanding which ones to change can be challenging. Some routes at some times of the year are more economical then others due to weather and ocean conditions but spotting this can be challenging on an individual passage level. Applying Data Analytics to ship performance measures (speed, fuel usage etc.) to ship geospatial and temporal data (location, time) with available third party data sources (weather, temperature, sea state etc.) would provide greater trend analysis and insight into the causes of vessel efficiency.

3.4 COATINGS

Below the waterline coatings to reduce fouling is well established as a means of maintaining vessel performance. Measuring this performance and monitoring overtime to is important to ensure that right decisions are made about when and which coatings to apply [1].

A Digital Twin tracking vessel performance, engine performance and weather conditions would provide operators an effective way of measuring how well their coatings are performing. With around 5-20% reductions in ship efficiency attributed to fouling, maintaining a focus on individual vessels is important to managing costs.

3.5 MAINTENANCE, REPAIR AND OVERHAUL (MRO)

The MRO of a component, equipment, system or ship can be a significant undertaking with requirements for trained personnel, spare parts, tools and technical information. Some sectors such as commercial aviation have very strict requirements and tight configuration control over technical data, drawings and standards which have enabled Digital Twin approaches to be applied. Within the Commercial Maritime sector, such practices are followed in some areas but not universally.

Digitisation of sources of information such as drawings, manuals, system information into a DT can greatly enhance and support the design, build and support of a ship through-life. Provision of timely and accurate information can enable more maintenance and repairs to be conducted successfully on-board without the need to resort to second or third line support possibly at addition cost to the operator [2].

Capturing the As-Built and As-Maintained through life Digital Twins of a platform is essential as reality can quickly deviate from the As-Designed plans for a ship or system. Laser scanning and transferring / processing 2D documents can enable the creation of a repository of useful information for maintenance teams.

Condition based monitoring of systems is another use of a Digital Twin to monitor the individual status of an item of equipment and potentially other information sources like flow rates into an engine etc. When running information is collected it is instantly available to provide a view on how the equipment is operating. By collecting and saving this information over periods of time Data Analytics can allow for trend analysis or recurrent patterns to be identified.

By tracking the performance of equipment under or over maintenance of equipment can be identified. Costs are therefore lowered by mitigating the costs of recovery from equipment failure caused by under maintenance but also avoiding ones where an overhaul is not needed at that time.

Greater trend analysis can then be applied by considering the operating context of the equipment or platform over its lifetime. By combining condition based monitoring with the ships context (location, weather etc.) insights such when best to overhaul a particular pump, on a particular ship, doing a specific route maybe a very different time to another ship on another typical route.

Digital Twins can also be beneficial in relaying support information back to experts ashore to make decisions. This is particularly beneficial should the technology be too complicated for the level of training of the crew at sea.

3.6 COMPLIANCE

Managing and monitoring compliance, whether this be from emissions to reporting requirements, benefits from a Digital Twin. DT can provide evidence from monitoring the configuration and usage of systems on ships to report compliance with local requirements. By having a digital ecosystem with information made easily available the overhead costs associated with regulatory audit can be made simpler and less costly.

3.7 TERMINALS AND SUPPLY CHAINS

Shipping companies are looking wider than just ships for transformations to cover the entire supply chain. By considering the whole supply chain from production by a customer to final delivery to the end customer shipping is an essential component but is only one step in the process [11]. The Digital Twin information is an enabler within the end-to-end supply chain operation that can be used to help the customer experience, identify opportunities for further process improvements, and reduce paper-based administration, as well as augmenting live asset tracking to provide more accurate and simpler query reporting for customers [3].

Greater access for customers and suppliers data to be shared with trusted partners is commonplace in other sectors and would transform the commercial shipping sector [4].

4. CHALLENGES TO IMPLEMENTING DIGITAL TWINS

4.1 TECHNICAL CHALLENGES

The technology required to implement Digital Twins on scale exists. Adopting it in organisations without an in depth understanding of the underlying technology can be a challenge.

4.1 (a) Sharing Data

Digital Twins are all about sharing data and to the widest audience possible whether this be on a ship, in an organisation or with an entire supply chain. Choosing, creating, migrating and maintaining a suitable data architecture can be a bewildering challenge. Setting out with a clear objective and adopting a services based architecture allows for data and services to be added later without a total redesign of the system.

As the lifetime of the Digital Twin and underlying asset (ship, system etc.) is likely to be far in excess of the software used to create the model, care must be taken to ensure the DT remains supportable and reusable in the future. Proprietary software represents a high risk of elements of the model becoming unreadable in the future [5].

Identifying all of the possible uses and applications of a Digital Twin is not possible. Instead, architects and engineers should define architectures that allows access to data from different sources and many data sources, with data features built in to respect business commercial sensitivities.

Communicating with the wider supply chain will present its own challenges as ship operator's work to protect their own and others Intellectual Property (IP). A technology that could facilitate this is block-chain as secure verification and validation of users is an inherent element of the technology.

4.1 (b) Cyber Security

Networking of systems, sharing information and use of automation all provide security vulnerabilities. Any Digital Twin platform needs to consider security from the initial design and deployment.

Digital Twins that expand beyond ships at sea will move the edge of the security boundary as well. Considering how to secure access but also safe guard critical functions like machinery control and navigation is an essential challenge.

4.1 (c) Connectivity

Communications and networking to other ships and ashore has been transformed by the introduction of geostationary satellite communications. In the next decade the introduction of CubeSat (small scale satellites) into Low Earth Orbit (LEO) and Medium Earth Orbit (MEO) is set to transform connectivity at sea again.

LEO based satellite communications will provide global coverage of high-speed data and communications including over the seas. Unlike geostationary systems, which are trained to stare at particular areas, these new constellations will be orbiting much closer to the Earth and so will appear to be moving rapidly overhead. To provide coverage over populated areas hundreds of satellites will be needed to provide coverage. For maritime platforms this will have the benefit of providing an affordable way of accessing high-speed internet connections anywhere at sea [7].

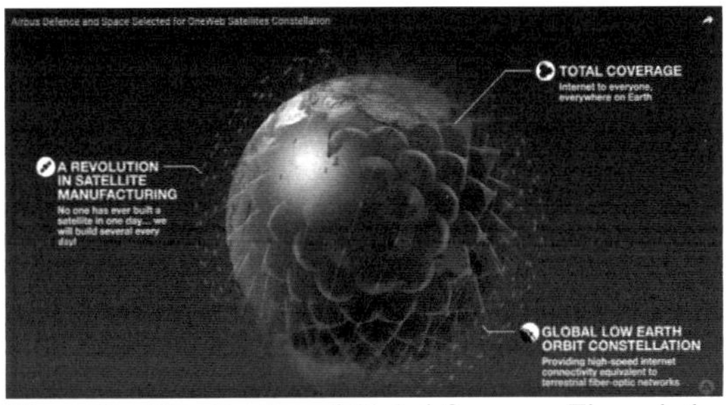

Figure 3 – Visualisation of Global Coverage Through the Use of LEO Satellite Communications

Additionally, the introduction of the 5G standard for communications ashore will be supported by satellite based 5G services.

Changes to 5G infrastructure and in particular aerial design will make the use of Active Electronically Scanned Arrays (AESA) more common. For ships wishing to make use of LEO satellite communications they will most likely need to use AESA planar or conformal planar arrays to track these fast moving satellites. Due to the apparent motion of the satellites, conventional, physically steered dishes will be unable to track a satellite and handover to the next one without breaks in transmission. AESA based technology allows for multiple links to be established concurrently at different points in the sky.

Communicating large amounts of data ashore will become more common place. With the move to Digital Twins ship operations, the need to ensure that ships are well connected will become increasing important for operators.

4.1 (d) Computing Platforms

The choice of computing platforms from massive on premises systems are being replaced often with Cloud Based Services. There are pros and cons to each option from security to available services to support further analytics. Managing commercial sensitivities and security constraints can be handled through Cloud computing although some commercial contracts may require on-premises solutions.

The right computing platforms can also reduce the effect of obsolescence by separating hardware from software management for a given service. A Software-As-A-Service (SAAS) type model can allow Ship Operators to gain access to the services they require but leave the SAAS operator to manage obsolescence, updates and upgrades including adding new features.

4.1 (e) Gaining Insights

Data and information alone do not improve the performance and efficiency of a system. It is important that the information and data gained from the digital twin can be turned into actionable and informative insights. This can be achieved primarily through closing the loop between the simulated environment and the measured real world environment, data analytics can be used to allow the user to track trends and outline areas for improvement.

Further iterations and improvements of digital twin should allow for the automatic diagnosis and detection of problems to remove the burden of the human operators. This functionality will require significant investment and historical data to allow machine learning algorithms to plot the key trends.

4.2 COMERCIAL CHALLENGES

As well as the technical challenges some of the most critical challenges are the commercial, legal and regulatory barriers.

4.2 (a) IP

Intellectual property and rights to the newly developed systems and technologies could be locked down and protected by those organisations that have invested in their development. However, due to the software centric nature of a DT system, it is also intrinsically difficult to protect. Companies working in this space will need to be conscious of IP limitations and considerations which could hold back or limit development.

4.2 (b) Access to Data

As previously discussed in section 4.1 (a) Digital Twins are all about sharing data and to the widest audience possible. Access to this data is essential and if this is restricted then it can affect the benefits of the entire concept. There is a requirement for choosing, creating, migrating and maintaining a suitable data architecture. If this architecture can be agreed, standardised and regulated then it would be a significant benefit to the entire sector.

4.2 (c) Including legacy platforms

It is not expected that these types of DT system can be implemented overnight, instead they will need to interact with a range of legacy platforms both within their own fleets and physically around them. It will be this time that will be the hardest test for DT, they will be under increased scrutiny looking for any problems or reasons not to invest in the systems.

One way to mitigate against this is the gradual adoption of a DT solution, starting with the information architecture required to enable DT, this can in turn be upgraded to include some elements of smart sensing and feedback before a full digital twin solution can be implemented. Using this key information architecture you can create ships that are DT ready without being fully DT enabled.

This information system will still have significant benefits in terms of support and maintenance planning, recovering any initial investment costs.

4.2 (d) Humans resisting change

With the introduction of new DT enabled vessels will create a significant change for those operating with the vessels. As with all change, there will be some people that will resist the change seeking to maintain their existing practices and ways of working. In order to counter this, change will need to be managed effectively from senior managers, introducing the benefits and a steady introduction of the physical change. By establishing well-documented practices and procedures and communicating these effectively the disruption can be minimised [11].

4.2 (e) Human training

In order to facilitate this change there will need to be a complete refresh of the human training for those working with the vessels. This new training will need to consider the human-autonomy teaming enabled by DTs and teach the users how to trust and maximise the potential of the DT systems.

4.2 (f) Effects on business models

With the benefits of increased efficiency and new opportunities also comes the threat of new business models. If managed carefully and successfully this could lead to prosperity for the company, however, if the company tries to maintain its old existing business models without any change it could quickly find itself outpriced and outpaced by its competitors. This is both an opportunity and a risk that needs to be carefully considered and managed.

7. CONCLUSIONS

The Maritime Industry 4.0 is expected to provide an increase in available computing power, improving connectivity at sea and the constant drive to manage costs is enabling new business functions whilst transforming existing ones. With greater access to operational information harvested from ships, ports, the oceans etc. ship operators can now gain actionable insights not previously possible. Digital Twins are the vehicle for this business improvement. By using and combining all the available data the application of data analytics, DTs can provide greater insights and value to businesses reducing unnecessary activities, minimising unexpected events and saving costs.

There are significant opportunities for DT technologies to improve efficiency with crew training, crew travel, fuel management and maintenance and repair schedules. In order to extract the full benefits of the digital twin system it must extend past the equipment and system level to cover the entire supply chain. These larger encompassing data and modelling environments will provide the information and connectivity required to run in the Maritime Industry 4.0.

There are however, still a large number of technical and commercial challenges to overcome to make a DT in maritime shipping a reality. A key enabler from the concept design stage is the digitisation of sources of information such as drawings, manuals and system information into a DT. It is this structure and information that can greatly enhance and support the design, build and support of a ship through-life. Provision of timely and accurate information can enable more maintenance and repairs to be conducted successfully on-board without the need to resort to second or third line support possibly at addition cost to the operator

It is not expected that the full scope of a DT for maritime shipping described in this report will become a reality overnight. Instead, it will be the result of iterations of concepts and vessels overtime. One thing that we do know, is that you cannot have the Maritime Industry 4.0 without a Digital Twin.

8. ACKNOWLEDGEMENTS

The authors would like to acknowledge all those who supported the digital twin research conducted at BMT and the production of this paper.

9. REFERENCES

1. Richard Meade, 'Total Cost of Operation: A Guide to Optimising Vessel Operating Costs', Lloyd's List, Sep 2018.

2. SSI, 'Managing Vessel Repairs with a "Digital Twin", https://www.maritime-executive.com/blog/managing-vessel-repairs-with-a-digital-twin, 4 Feb 2019.

3. A.P. Moller – Maersk, 'Annual Report 2018', https://investor.maersk.com/news-releases/news-release-details/annual-report-2018m, 21 Feb 2019.

4. K. D. Adamson, 'Shipping and 800lb Gorilla', FN Books, 2017.

5. Laurence Goasduff, 'Confront Key Challenges to Boost Digital Twin Success', https://www.gartner.com/smarterwithgartner/confront-key-challenges-to-boost-digital-twin-success/, 13 Mar 2018.

6. A. Parrott, Dr. L. Warshaw, 'Industry 4.0 and the digital twin', https://www2.deloitte.com/content/dam/Deloitte/cn/Documents/cip/deloitte-cn-cip-industry-4-0-digital-twin-technology-en-171215.pdf, 2017.

7. I del Portillo, B Cameron and E Crawley, "A Technical Comparison of Three Low Earth Orbit Satellite Constellation Systems to Provide Global Broadband", 69th International Astronautical Congress 2018, 1 Oct 2018.

8. Jonathon Scott, "How does one get started with PLM and the Digital Twin", Machine Design, https://www.machinedesign.com/cad/how-does-one-get-started-plm-and-digital-twin

9. BCG, 1 BCG, Digital in Engineering and Construction, McKinsey-Construction Productivity, 2016.

10. IMO Briefing, IMO sets 2020 date for ships to comply with low sulphur fuel oil requirement, http://www.imo.org/en/MediaCentre/PressBriefings/Pages/MEPC-70-2020sulphur.aspx, 28 Oct 2016.

11. Goasduff, L. (2019). Confront Key Challenges to Boost Digital Twin Success. [online] Gartner.com. Available at: https://www.gartner.com/smarterwithgartner/confront-key-challenges-to-boost-digital-twin-success/ [Accessed 23 Sep. 2019].

9. AUTHORS BIOGRAPHY

Simon Willmore holds the current position of Business Unit Lead at BMT. He is responsible for oversight of the technology related capabilities support ships and other engineering areas. His previous experience includes a career in the Royal Navy as a Weapon Engineering Officer and as part of BMT has supported multiple ship design and support projects.

Gren Mercer holds the current position of Principal Systems Engineer at BMT. Gren has wide domain knowledge of technical engineering activities within defence, industry and research organisations with over 35 years of experience. In this time he has applied systems thinking to maritime and land defence projects covering concept studies and operational analysis, requirements development, and some relevant technical specialist areas such as electronics and software development.

Jake Rigby holds the current position of Research and Development Lead at BMT. He is responsible for Research Portfolio Management, Academic Engagement and Horizon Scanning. His previous experience includes 4 years as a Naval Architect working on a range of ship designs and consultancy projects.

CLASSIFYING BOW ENTRY EVENTS OF WAVE PIERCING CATAMARANS IN RANDOM WAVES USING UNSUPERVISED AND SUPERVISED TECHNIQUES

B. Shabani, **J. Lavroff**, **D. S. Holloway**, University of Tasmania, Australia

S. Penev, UNSW Sydney, Australia

D. Dessi, Italian National Research Council, Italy

G. Thomas, University College London, UK

SUMMARY

An onboard monitoring system can measure features such as stress cycles counts and provide warnings due to slamming. Considering current technology trends there is the opportunity of incorporating machine learning methods into monitoring systems. A hull monitoring system has been developed and installed on a 111 m wave piercing catamaran (Hull 091) to remotely monitor the ship kinematics and hull structural responses. Parallel to that, an existing dataset of a geometrically similar vessel (Hull 061) was analysed using unsupervised and supervised learning models; these were found to be beneficial for the classification of bow entry events according to the kinematic parameters. A comparison of different algorithms including linear support vector machines, naïve Bayes and decision tree for the bow entry classification were conducted. In addition, using empirical probability distributions, the likelihood of wet-deck slamming was estimated given vertical bow acceleration thresholds.

1. INTRODUCTION

Wet-deck slamming (wave impact against the cross-deck structure) is an important consideration in the structural design of catamarans [1]. The entry of the centre bow (shown in Figures 1 and 2) in waves minimises the risk of deck-diving and improves seakeeping characteristics but adds complexity and non-linearity in the hull-wave interaction during arch filling and wet-deck slamming [2-5]. Large transient slam loads may occur, generating whipping and structural vibrations [6] and in the long term, these vibrations can cause fatigue damage [7]. Wet-deck slam loads on catamarans can also generate extreme loads with the potential to damage the vessel structure as shown in Figure 3.

The severity of slam forces on high speed wave piercing catamarans (WPCs) has been reported to be in the order of the vessel weight but a high level of uncertainty still exists in determining these loads. There have been numerical simulations of motions and loads of WPCs [8] and model test programmes for measuring the extreme model scale loads with respect to various speeds and wave heights in both regular and irregular waves [9, 10]. Recent studies on motions and loads of WPCs have been based on hydroelastic segmented model tests mainly conducted in regular [1, 2, 4, 5, 9, 11] and irregular [6, 10, 12] head waves.

Figure 1: HSV-2 SWIFT, a 98-meter wave piercing catamaran built in 2003 (Hull 061) (Incat.com.au, 2019).

Figure 2: Volcan De Tagoro, A 111- m wave piercing catamaran built in 2019 (Hull 091) (Incat.com.au, 2019).

Figure 3: Side shell buckling resulting from wet-deck slam loading on a 96-m vessel built in 1998. [4].

Slam loads, pressures and kinematics of WPCs during bow entry events in regular head seas were investigated previously in [3-5, 13-15]. In regular waves, it was seen that wet-deck slamming occurs when the bow immersion depth relative to undisturbed water surface reaches a certain limit which is a function of wave encounter frequency. This is shown in Figure 4 for two dimensionless wave encounter frequencies. In real operating conditions, a large set of data can lead to more accurate predictions of wet-deck slamming events. This is important from a structural design perspective considering that the existing prediction methods and design rules for WPCs are yet to be developed [7].

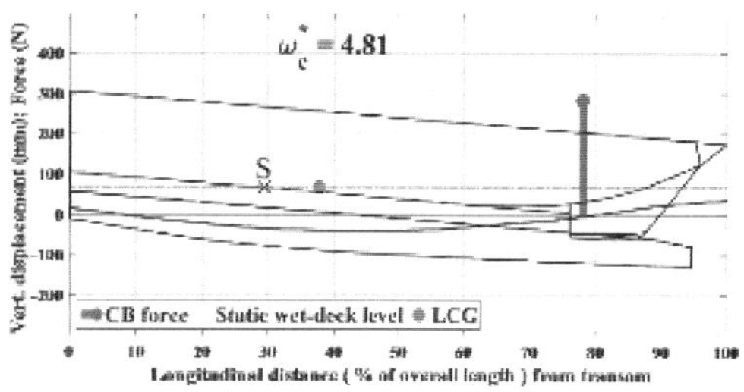

(a)

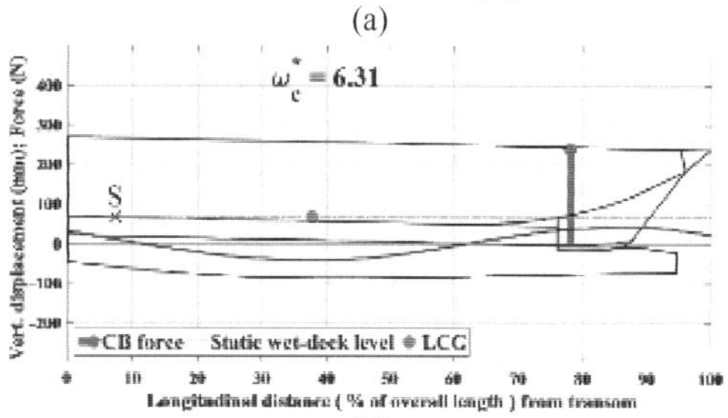

(b)

Figure 4: Vertical model displacement, encountered wave profile and CB load at slamming instants for HSM02 model with the parent CB in $\boldsymbol{h_w} = \mathbf{90}$ mm, $\boldsymbol{V_m} = \mathbf{2.89}$ m/s for two dimensionless wave encounter frequencies ($\boldsymbol{\omega_e^*}$). 'S' shows intersection point of the static wet-deck level and instantaneous wet-deck line [3].

A hull monitoring system, which integrates strain gauges, accelerometers and other sensors, can provide valuable information for both ship design and operations. A monitoring system can have standard features specified in class guidelines such as stress cycles counts, warnings due to slamming and excessive motions. Considering the current technology trends and digitalisation affecting services, products and processes, smart and connected hull monitoring systems incorporating machine learning (ML) and deep learning methods are of interest.

When it comes to slamming, a hull monitoring system could be developed in order to satisfy multiple requirements. For example, it may have certain features to (1) predict ahead of time if the vessel is likely to be subjected to wave impacts given the current operating conditions (2) provide awareness of the likelihood and severity of slamming events that could hypothetically lead to either local or global structural damage (3) propose a modified speed or change of course to minimise the structural risks (4) automatically identify slamming events and provide a statistical summary of slamming occurrences and severity according to slam-induced accelerations and hull stresses. The first three features mentioned above represent some of the problems that were addressed by Ochi and Motter [16] through a closed-form statistical model. More details of the Ochi slamming conditions and slamming probability can be found in [17] and a review of slamming identification methods can be also found in [18]. Nevertheless, the alternative methods based on ML have yet to be developed to identify and classify slamming events in random waves.

Various ML techniques [19-23] have been developed to address a variety of classification problems. Among a broad range of applications, ML models for structural health monitoring, damage detection and predictive maintenance of mechanical systems, ships and offshore structures [24] can be highlighted.

In this paper, the development of a remote monitoring system will be presented, and machine learning algorithms will be applied to classify centre bow entry events in random seas. This is part of a broader research project which aims to develop smart and connected hull monitoring systems specific to WPCs.

2. FULL-SCALE DATASETS

2.1 HULL 091 MONITORING DATA

A hull monitoring system has been developed for Hull 091 Incat vessel (see Figure 2). The vessel was instrumented in July 2019 using a motion reference unit (MRU), a bow accelerometer and 10 strain gauges. An ultrasonic wave sensor was placed in the bow area to measure the incident wave profiles. An overview of the remote monitoring system is shown in Table 1. Figure 5 shows the approximate longitudinal locations of the cDAQ, laptop, MRU, strain gauges, bow accelerometer and the ultrasonic sensor. The sampling rates were not consistent across the sensors because of the requirements of modules/sensors. The sampling rates were set to 1000 Hz for strain gauge data, 100 Hz for the bow and MRU acceleration data, 50 Hz for MRU heave, pitch and roll data and 9 Hz for the ultrasonic sensor.

A LabVIEW program was developed to record and upload data automatically to a Google Drive account given a set of predefined rules and triggers. For instance, the LabVIEW program records full raw data as a technical data management streaming (TDMS) file and creates a subset file when the bow vertical acceleration is above 0.5*g* with a window of 30 seconds. In addition, the rain flow counting algorithm available in LabVIEW was included to output a comma-separated values (CSV) file, providing a summary of strain cycle counts from each strain gauge.

Table 1 Overview of Hull 091 remote monitoring components

Category	Item	Quantity	Description
Sensors /data source	Strain gauges	10	HBM 1-LY43-6/350
	Accelerometer (3-axial)	1	CrossBow- CXL04GP3
	Ultrasonic sensor	1	ToughSonic 50(TSPC-21SRM-485)
	Motion reference unit (MRU)	1	SBG Systems- Ellipse2-A
	Global Positioning System (GPS) receiver	1	Hull 091 GPS distributor
Data acquisition	Data acquisition (DAQ) module	1	National Instrument (NI cDAQ-9174)
	Strain gauges module (8 channel)	2	National Instrument (NI-9236)
	Universal Input module (4 channel)	1	National Instrument (NI-9219)
Computer, accessories and software	Laptop	1	Dell latitude 7490
	USB Hub	1	Powered USB hub
	Onboard Monitoring Software	1	Customised LabVIEW program
	Remote Desktop Access	1	TeamViewer
	Laptop tray	1	RAM Universal tray (RAM-234-3)
	Cabinet	1	PCLocs – Carrier 10
	NMEA to USB convertor	1	Digital Yacht
Storage & connectivity	External hard drive	1	Samsung- USB-C 1T SSD
	Cloud based storage	100GB (scalable)	Google Drive
	WiFi/LAN router	1	Digital Yacht 4G Connect PRO
	4G antennas	2	Digital Yacht 4G Connect PRO
	Sim Card	1	Simyo 4G

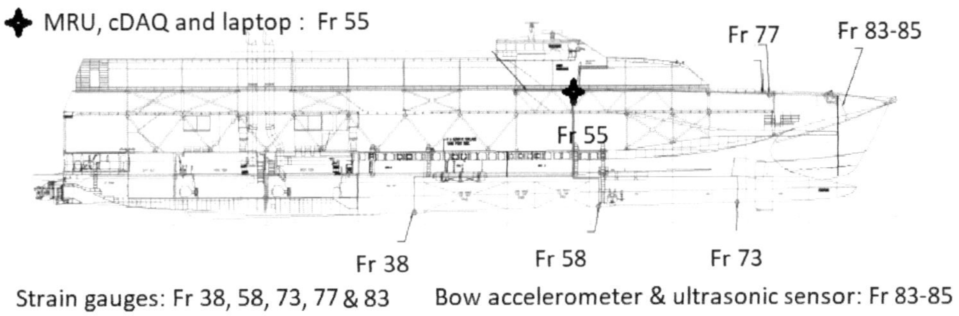

Figure 5: Hull 091 cabling diagram from Frame 55 to frames (Fr) 38, 58, 73, 77, 83 and 85, showing the approximate longitudinal locations of the cDAQ, laptop, MRU, strain gauges, bow accelerometer and ultrasonic sensor listed in Table 1.

It is worth mentioning that developing a customised LabVIEW program enabled the integration of the MRU and GPS data. This was achieved by incorporating SBG Systems (Ellipse2-A) and NMEA-0183 GPS drivers, to the strain gauges, bow accelerometer and ultrasonic sensor data acquired from NI 9236 and 9219 modules. Furthermore, a file management function was included so that the files could be uploaded into the Google Drive and deleted from the computer hard drive. The choice of Google Drive here is due to having a general purpose, temporary storage for preliminary data access and some other user-friendly features such as automatic synchronisation and backup to manage large data files. The customised upload code developed in the LabVIEW program can be adjusted to upload the data into a cloud platform which enables advanced features such as machine learning, deep learning and near real-time monitoring. Such features, for example, are available in public cloud computing services such as Amazing Web Services (AWS) Google Cloud Platform (GCP), and Microsoft Azure. Figure 7 shows a development process of the hull monitoring system which incorporates cloud computing services that will be developed in future stages.

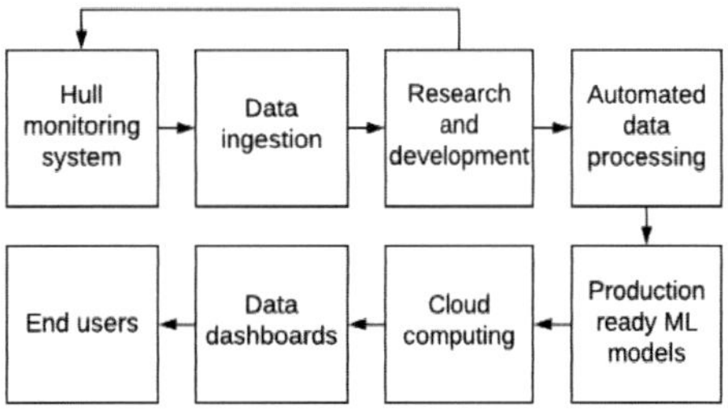

Figure 6: The development process of the hull monitoring system.

The delivery voyage of Hull 091 from Hobart, Tasmania to Canary Islands, Spain took place between 15 July and 15 August 2019. In the absence of Hull 091 data being fully analysed at the time of writing this document, Hull 061 sea trials data was used in order to develop a classification algorithm for the bow entry of high-speed WPCs. Table 2 compares main particulars of Hull 061 and Hull 091.

Table 2 Main particulars of Hull 091 and Hull 061

	Hull 091	Hull 061
Length overall	111.9 m	97.22 m
Length waterline	103.2 m	92 m
Beam Overall	30.5	26.6 m
Draft	4.1 m	3.434 m
Demihull beam	5.8 m	4.5 m
Max Deadweight	1000 tonnes	670 tonnes
Speed	42.4 knots	38 knots

2.2 DEVELOPMENT OF ML PIPELINE USING EXISTING HULL 061 DATA

An existing dataset (Hull 061 dataset) was used to propose a possible architecture for an ML pipeline, in which ML workflows can be automated as described in Figure 7. The use of Hull 061 dataset was an important step given the fact that the two vessels are geometrically similar, noting that the main objective of the ML workflows in this work is to classify the bow entry events according to the kinematics of centre bow entry [3] in random waves. In addition, the successful instrumentation of Hull 061 and lessons learnt through that process were the key for developing an ongoing monitoring system for Hull 091. The choice of having an ML model developed for classifying bow entry events of Hull 091 in the present work is also linked to Hull 061 bow entry events which resulted in wet-deck slamming occurrences [25, 26] in various speeds and wave heights during HSV-2 Swift sea trials.

The seakeeping and structural loads trials of Hull 061 were conducted in May 2004. The vessel was extensively instrumented by 47 strain gauges and 4 triaxial accelerometers located at the bow, bridge, LCG, and flight deck. The roll, pitch and yaw were measured at the longitudinal centre of gravity and a Tsurumi Seiki Co. Ltd (TSK) wave height meter system was installed at the bow (see Figure 7). Several parameters of the shipboard control systems were also monitored including the position of the T-foil and trim tabs, waterjet nozzle angle and waterjet shaft speed. Table 3 shows an overview of Hull 061 seakeeping and structural loads data obtained from sea trials between 11th and 17th of May 2004. The accumulated data rows from all runs were approximately 24 million rows, an equivalent of 66.6 hours of logged data at 100 Hz.

The sea trials were conducted in various wave heights and different octagons, each with five legs from head seas to following seas at a constant ship speed throughout each octagon. Since head seas accounted for a larger number of slams in comparison to other headings [25], a total of 9 runs in head seas were selected for current analyses as listed in Table 4.

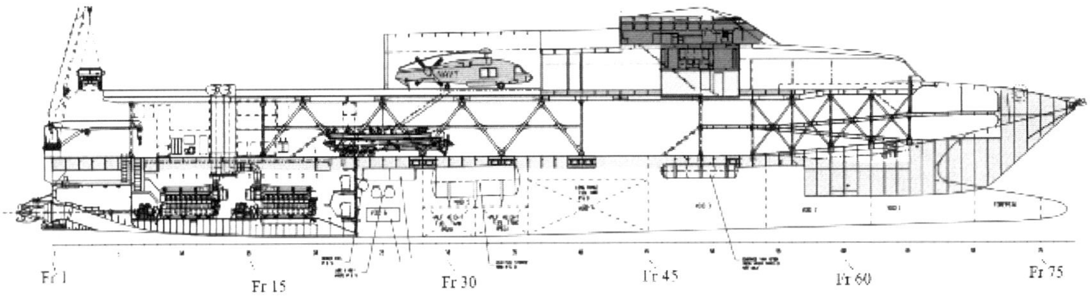

Figure 7: General arrangement (GA) of Hull 61. During sea trials between 11th and 17th of May 2004, the vessel was instrumented by several strain gauges and sensors from which 16 global strain gauges were located at frames (Fr) 6,12, 23, 25, 31 , 46, 61, 65, 67, four accelerometers were located at the bow, under the bridge console, LCG, and under the flight deck, and at a TSK wave height meter was located at Fr 72.

Table 3 An overview of HSV-2 Swift (Hull 61) seakeeping and structural loads data obtained from sea trials between 11th and 17th of May 2004

Total Runs	Sea State	Significant Wave Height (m)	Ship Speed (knots)	Number of channels	Sampling rate	Total data rows
159	4-5	1.8- 3.8	10-38	79	100 Hz	24 million *

Table 4 Selected runs in head seas for Hull 061 data series

Run number	Relative heading	Significant wave height (m)	Modal period (s)	Speed (knots)	Ride control status	Run duration (min.)
70	Head seas (all selected runs)	2.4	8.2	20	ON	21.1
92		3.0	11	30	ON	22.2
99		2.0	7.5	35	OFF	19.4
145		2.4	10.2	35	ON	19.5
159		1.7	8.4	30	OFF	20.2
174		2.9	10.2	35	ON	20.4
180		2.8	9.7	35	OFF	20.3
192		1.9	7.6	30	ON	19.7
206		1.6	7.2	15	OFF	21.1

3. DATA ANALYSES APPROACH

3.1 TIME SERIES SEGMENTATION & FEATURE EXTRACTION

The rate of wet-deck slamming occurrence is a function of the centre bow and wet-deck geometry, speed, wave height and other operational factors [25], but the condition in which a wet-deck slam can occur is often simplified by considering the relative vertical bow displacements along the centre bow [3, 27]. The rate of wet-deck slamming occurrence therefore can be described by the number of bow entries exceeding a threshold value [27], noting that the variability of the threshold relative vertical bow displacement can be included [27].

A new approach is proposed in this work with the objective of having an automated data pipeline for describing bow entry events, in particular those that are likely associated with wet-deck slamming events.

Figure 8 shows the overall approach proposed for classifying the bow entry events that incorporates unsupervised and supervised classification methods [23].

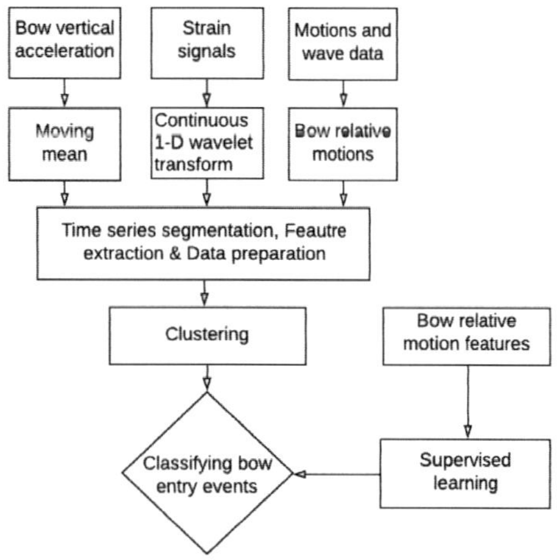

Figure 8: The process of centre bow entry classification.

As mentioned earlier, various methods for identification of slamming events have been proposed previously including whipping-based criterion [28], maximum rate of change of stress [18, 25], and wavelet methods [29]. The approach proposed in Figure 8 considers bow vertical accelerations, strain gauge measurements and relative motions as the inputs for an unsupervised learning algorithm for the classification of the bow entry events. The unsupervised model can be used for onboard or cloud-based analyses when other all inputs are available from the measurements. In addition, a supervised learning model is proposed which only requires the relative motions data to classify the bow entry events. The supervised model is then an alternate solution for simulations of wet-deck slamming analyses in random waves according to the bow entry kinematics.

As shown in Figure 8, in the first step, the moving mean of the bow vertical acceleration was used to divide the time series into segments so that each segment represented a time window between the zero crossings of the moving mean signal (see Figure 9). Only segments with positive moving mean accelerations (see for example Figure 9 between 10.3 and 10.35 minutes) were selected for the analyses as they are linked to the centre bow entry events.

In a parallel step, by parsing strain signals through the 1-D continuous wavelet transform (CWT) function in MATLAB version 2019a (by The MathWorks, Inc.), the frequencies with the highest magnitudes were identified for each strain signal within each segment. The wavelet transform is obtained using the analytic Morse wavelet, with $L1$ normalisation, so equal amplitude oscillatory components in the data at different scales have equal magnitude in the CWT function.

Figure 11 shows an example of the wavelet transform applied to a standardised strain signal, which is scaled to a zero mean and standard deviation of 1. The strain signal was low-pass filtered by a cutoff frequency of 5 Hz before the wavelet transform. Frequencies with the highest magnitude are presented as a function of time, showing a peak frequency of 2.83 Hz which plateaued around a time of 14.72 minutes from the start of recording, noting that this was in a segment at which a wet-deck slamming occurred with a peak acceleration of above 2 g.

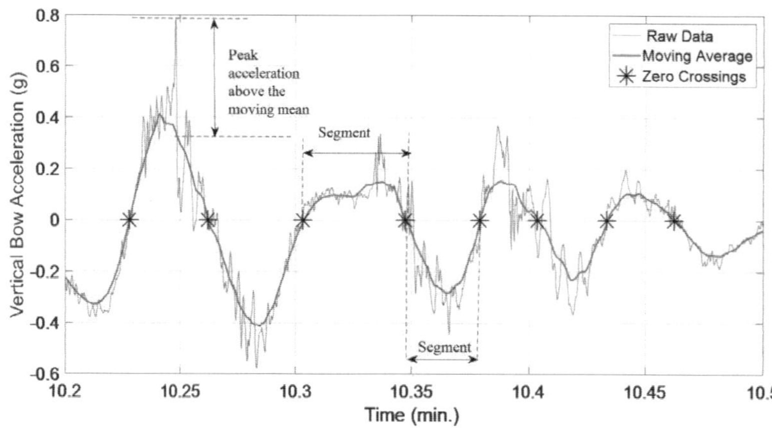

Figure 9: A time record of vertical bow acceleration and its moving average, showing a typical peak acceleration above the mean, and two segments with positive and negative moving mean accelerations. Raw data is taken from Run 145 at a speed of 35 knots in head seas with a significant wave height of 2.4 m.

(a)

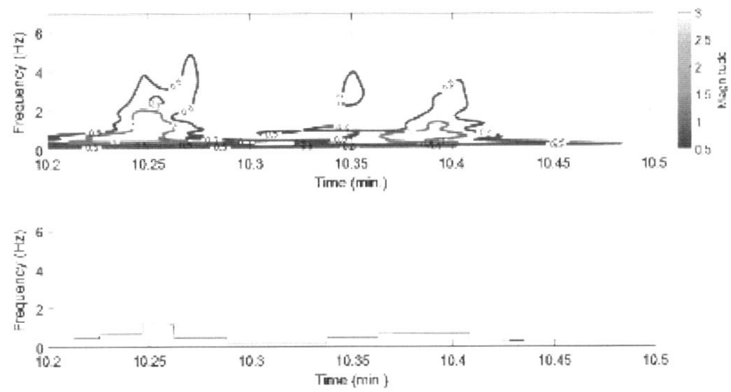

(b)

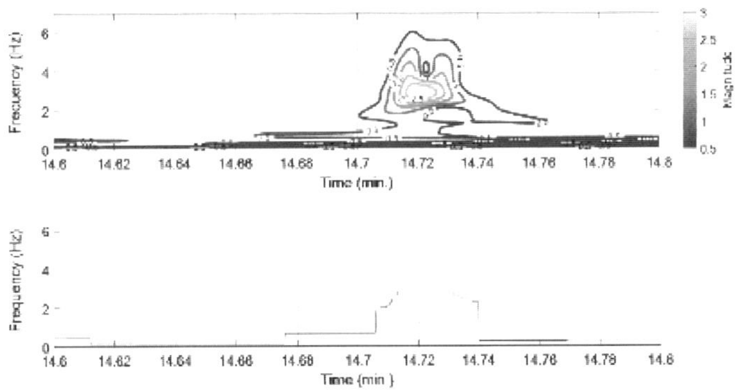

Figure 10: 1-D wavelet transform's frequency and magnitude of a standardised strain signal (contour plots), and frequencies with the highest magnitude as a function of time (line plots): (a) between 10.2 and 10.5 min., corresponding to peak accelerations shown in Figure 10; (b) between 14.6 and 14.8 min, corresponding to an impact with a peak acceleration of above 2g. Raw data is taken from strain gauge T1_10 at Fr 61 measured during Run 145 at a speed of 35 knots in head seas with a significant wave height of 2.4 m. The wavelet transform is obtained using the analytic Morse wavelet. The color bar shows the magnitude of oscillatory components with $\boldsymbol{L1}$ normalisation in MATLAB 2019a.

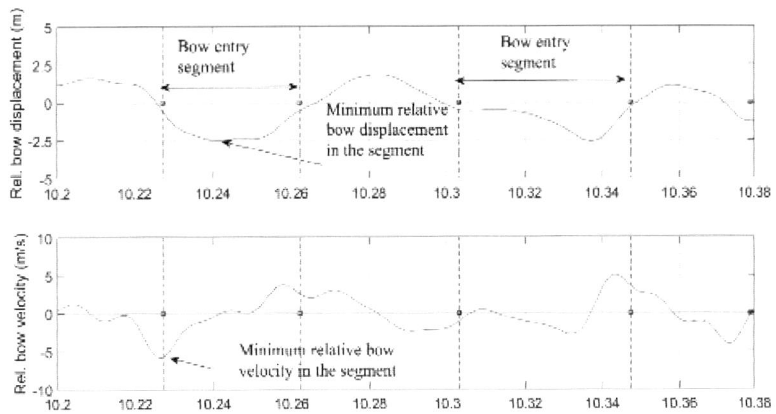

Figure 11: Representation of (a) minimum relative bow displacement in bow entry segments (b) minimum relative bow velocity in bow entry segments. Raw data is taken from Run 145 at a speed of 35 knots in head seas with a significant wave height of 2.4 m.

Table 5 Variables selected for the classification of the centre bow entry

Variable Names	Data category	Selected features	Presentation	Standardisation
var1	Moving average of bow vertical acceleration	Maximum value for each segment	Figure 9	Yes
var2	Vertical bow acceleration above the moving average	Peak value for each segment	Figure 9	Yes
var3	Peak frequencies obtained from the wavelet analyses	Maximum frequency value for each segment	Figure 10	Yes
var4	Peak magnitudes obtained from the wavelet analyses	Maximum magnitude for each segment	Figure 10	Yes
var5	Relative bow displacement	Minimum value for each segment	Figure 11	Yes
var6	Relative bow velocity	Minimum value for each segment	Figure 11	Yes

The centre bow entry in waves was analysed by considering the vertical displacement of the centre bow relative to the measured wave elevation in the bow area. Two parameters were selected to describe the centre bow entry in each segment: (1) relative bow displacement (2) relative bow velocity, as shown in Figure 11. The segment's boundaries in the figure are zero crossing points found by the analyses of the moving average of the vertical bow acceleration as shown in Figure 9.

Table 5 shows a summary of features extracted for each segment. Features are either maximum or minimum values of quantities described in Figures 9-11.

In the data preparation step before clustering, features were scaled to have a mean value of zero and a standard deviation of 1. The standardisation function is $Z = \frac{x-\bar{X}}{S}$, where x is sample data, $\bar{X}$ is mean and S is standard deviation of the sample data. It is worth noting that the standardisation was conducted after the features listed in Table 5 were calculated from all runs listed in Table 4. The normalisation is required for the subsequent data clustering (described in the following section) so that the clustering is not biased towards any particular feature.

3.2 DATA CLUSTERING AND SUPERVISED CLASSIFICATION

Data clustering or unsupervised classification is a technique that does not require structured prior information about groups or classes in a given dataset [30]. The objective is to find natural groups (i.e. clusters) within the dataset based on a similarity measure such that each cluster represents a meaningful category according to a selection of variables or features extracted from the original dataset. Various algorithms have been developed for clustering including hierarchical clustering, partitioning and density-based partitioning algorithms [31]. For instance, in the *k*-means algorithm, which is one of the most widely used methods in partitioning, the objective is to find *k* clusters from *n* observations so that each observation can be assigned to a cluster according to the minimum distance (e.g. Euclidian metric) from the mean value of each cluster [19, 20]. While there are various forms of *k*-means, the main steps are [20, 32]:

Step1. Select an initial partition with *k* clusters; repeat steps 2 and 3 until cluster membership stabilises.

Step 2. Generate a new partition by assigning each pattern to its closest cluster center.

Step 3. Compute new cluster centers

In the steps above the number of clusters (*k*), distance metric and initialisation of clusters are user specified parameters. Use of the Euclidean distance as the distance metric is a typical choice; other metrics include Manhattan, Chebyshev and Minkowski metrics [33]. The cluster initialization error can be also minimised through different approaches [34, 35]. It is worth noting that an extension of the basic *k*-means is fuzzy *c*-means [36, 37], in which each observation can be assigned to multiple clusters. An in-depth review of the application of clustering techniques can be found in [19], noting that the paper is focused on time series clustering which is the case for the present study.

The clustering analyses were conducted using the statistics and machine learning toolbox available in MATLAB version 2019a, in which a function has been developed that uses a *k*-means clustering method referred to as Lloyd's algorithm [38], but the default setting of the function uses *k*-means++ [39]. The *k*-means++ algorithm outperforms Lloyd's algorithm in speed and accuracy by incorporating a different seeding technique that creates *k* clusters one by one according to a probability function, as opposed to an initial partitioning with *k* clusters in the first step.

In supervised learning the objective is to train a classifier based on a set of labeled data or training examples. Various classifiers have been developed for this task including Bayesian classifiers, nearest neighbor classifiers, linear and polynomial classifiers, artificial neural networks and decision trees [21]. A review of classification techniques and algorithms with a focus on Internet of Things and sensory data analysis can be found in [22]. In this work, the classification learner application in MATLAB version 2019a was used to train classifiers based on relative motion features.

4. RESULTS

4.1 CLASSIFICATION OF BOW ENTRY EVENTS USING UNSUPERVISED LEARNING

The centre bow entry events were analysed from a total of approximately 3 hours of measurements accumulated from 9 runs in head seas with a significant wave height in the range between 1.6 m and 3.0 m, as presented in Table 4, in order to classify the events into 3 groups with respect to 6 features listed in Table 5 showing a set of features obtained from strain measurements, vertical bow acceleration and bow relative motion data. The clustering was achieved by using *k*-means++ algorithm, which was briefly introduced in the previous section. As mentioned earlier, the number of clusters (k) is a user specified parameter and the choice of k=3 here can be altered to a higher or lower number. Bow entry events with a minimum relative bow displacement of - 0.5 m at the reference frame of relative motion measurement (i.e. Fr72-Hull 061) were selected. This resulted in a total of 2378 bow entry events, from which 58%, 31% and 11% were identified as "group 1", "group 2" and "group 3", respectively. The differences between the groups can be explained by box plot presentations of feature variables in each group as shown in Figures 12-14. The line inside each box is the median calculated for each distribution, while the upper and lower edges of each box present the 25th and 75th percentiles, respectively. The most extreme data points shown by the upper and lower whiskers correspond approximately to $\pm 2.7\sigma$ if the data were normally distributed, where σ is the standard deviation of sample population. The data points outside these limits (i.e. 99.3% coverage) are considered as "outliers" and are shown by separate data points in each case.

It can be inferred from Figure 12(b) that group 3 contains strong wet-deck slamming events given the distribution of the outliers in comparison to that in groups 1 and 2. In addition, the likelihood of wet-deck slamming events to be in groups 1 and 2 is much lower than that in group 3, supported by the frequency distributions presented in Figure 13 (a) in which groups 1 and 2 data are almost out of the range of expected whipping frequency (i.e. 2 -3 Hz) for Hull 061.

For the above reasons, it can be argued that the kinematics conditions for wet-deck slamming occurrences in terms of bow relative displacement and velocity are best described by group 3 distributions plotted in Figure 14. However, it is difficult to draw a conclusion for conditions leading to wet-deck slamming occurrences based on the individual distribution of either minimum relative bow displacement or relative bow velocity. For example, the top whiskers in groups 2 and 3 show little differences in Figure 14, indicating group 2 bow entry events could be easily misclassified as group 3 or vice versa. This is an important consideration as minimum relative bow displacement is often considered as the only parameter in the simulation of wet-deck slamming occurrences of WPCs in a seaway Such an approach can be, to a certain degree, misleading in terms of the rate of slamming occurrences if only one reference section is used.

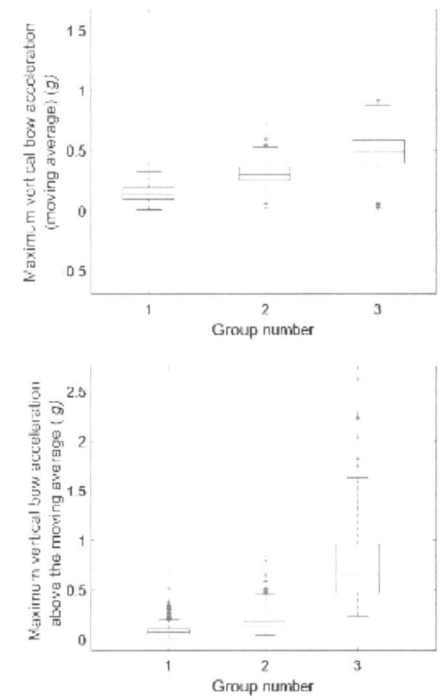

Figure 12: Box plot presentations of (a/top) maximum values of moving average of vertical bow acceleration during bow entry (*var1*) (b/bottom) maximum vertical bow acceleration above the moving mean (*var2*). Groups 1, 2 and 3 were determined by k-means ++ algorithm with input variables shown in Table 5.

Figure 15 shows bow entry event rates for various runs at different speeds and significant wave heights. As can be seen the bow entry event rates in group 2 can be several times higher than that in group 3. Interestingly, the results in Figure 15 show that in 30 and 20 knots the rates of group 3 bow entry events are higher than that in 35 knots, but such events are eliminated at a speed of 15 knots, highlighting the importance of effective speed reduction for decreasing the rate or minimising the likelihood of wet-deck slamming occurrences from operational perspective.

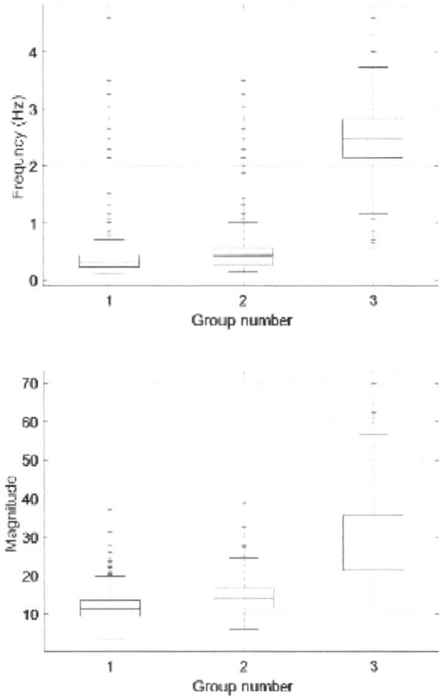

Figure 13: Box plot presentations of (a/top) frequency with maximum magnitude in the wavelet transform during bow entry (*var3* in Table 5) and (b/bottom) wavelet transform maximum magnitude during bow entry (*var4* in Table 5), where raw data obtained from a strain gauge at Fr 61. Groups 1, 2 and 3 were determined by k-means ++ algorithm with input variables shown in Table 5.

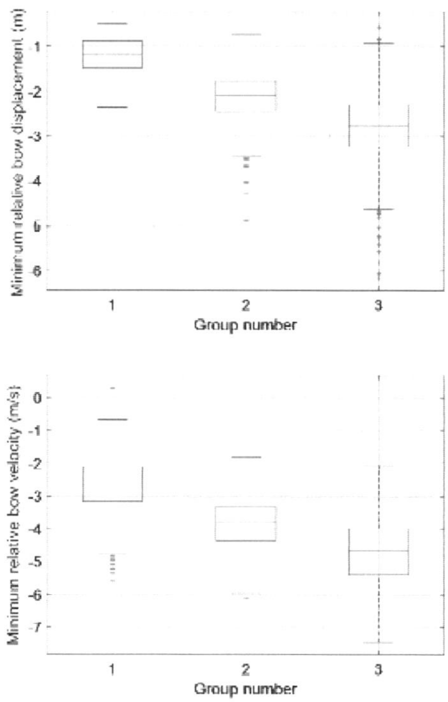

Figure 14: Box plot presentations of (a/top) minimum relative bow displacement (var5 in Table 5) (b/bottom) minimum relative bow velocity (var6 in Table 5). Groups 1, 2 and 3 were determined by k-means ++ algorithm with input variables shown in Table 5.

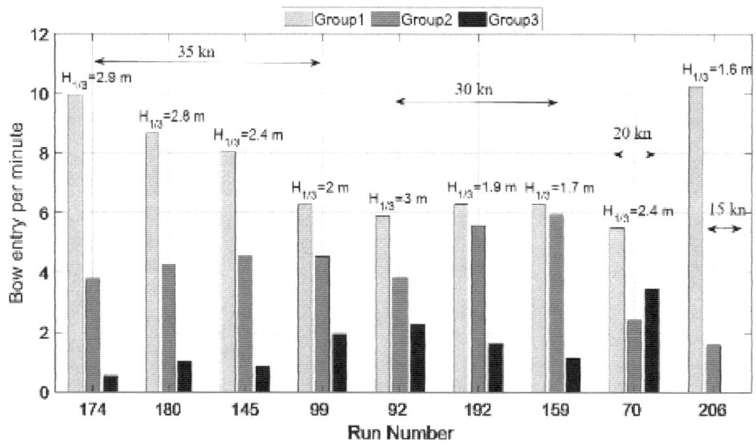

Figure 15 The rate of bow entry per minute for various groups for each individual run.

4.2 CLASSIFICATION OF BOW ENTRY EVENTS USING SUPERVISED LEARNING

Considering minimum relative bow displacement and velocity as a pair to classify bow entry events may lead to a better estimation of group 3 bow entry events in comparison to a single parameter criterion such as threshold relative vertical bow displacement. This is in fact a common approach for evaluating bottom slamming occurrences and the probability of such events in monohulls [17]. However, it should be noted that the conditions in which bottom entry events are classified for monohulls as either "slamming" or "no slamming" are usually defined by certain kinematics rules [16] rather than by a data driven approach taken in the present work.

One approach for the classification of centre bow entry events could be to calculate distances from the centroid of each group shown in Figure 16 to determine the type of bow entry events. More broadly, supervised learning classification algorithms such as support vector machines (SVMs), naïve Bayes or decision trees can be applied. The full explanation of supervised learning models and their applications for the classification of the bow entry events is beyond the scope of the present work but it is of interest to show how these models would generalise the bow entry classification problem by considering bow entry groups, shown in Figure 16, as training examples. An overview of possible outcomes from these models is presented in Figure 17. Table 6 shows the confusion matrix and the accuracy of these models with MATLAB's default settings (version 2019a) for each algorithm.

Each algorithm resulted in a different pattern although the data used for training were identical. Figure 17(c) shows the result of a coarse decision tree which indicates a minimum relative displacement and velocity of about -2.6 m and -3 m/s for group 3 bow entry events, where the former value is comparable to the wet-deck clearance of Hull 061 in calm water and the latter is about 15% of a full speed of 38 knots. In contrast, at a relative displacement of -2.6 m, the two other models (i.e. linear SVM and Gaussian naïve Bayes) shown in Figure 17 (a & b), suggest a relative bow velocity of - 5.3 m/s although these models describe the bow entry events quite differently in other regions.

Moreover, the number of strong wet-deck slamming events in each group was found to be very different to that seen in the original groups used for training. This is shown in Figure 18 which shows the maximum bow vertical acceleration above the moving mean, or feature defined as *var2*, for each group for the three classifiers. As can be seen in Figure 18 (a-c), there are many outliers suggesting wet-deck slamming events in groups 1 and 2 as opposed to that seen in Figure 12 (b). Thus, it is difficult to judge whether wet-deck slamming events are best described by these classifiers without prior knowledge about the distribution of wet-deck slamming events in each group. Therefore, a probabilistic description of wet-deck slamming events is required.

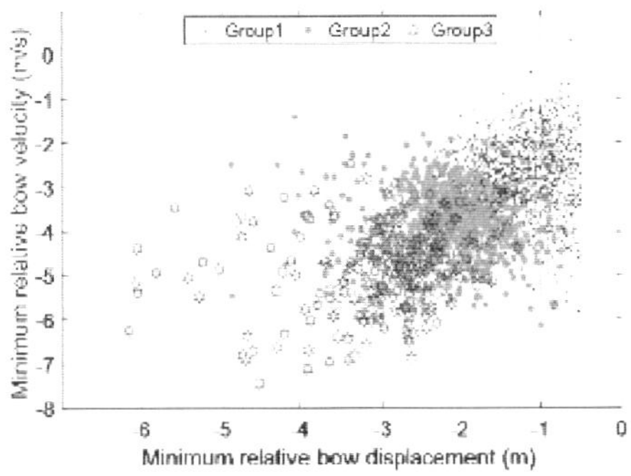

Figure 16 Minimum relative bow displacement versus minimum relative bow velocity during bow entry events in groups 1, 2 and 3.

4.3 PROBABILITY OF SLAMMING

The probability of wet-deck slamming $P(\text{slam})$ can be estimated as:

$$P(\text{slam}) = \frac{N_s}{N_{be}} = \frac{\sum_{i=1}^{k} N_{S,i}}{\sum_{i=1}^{k} N_{be,i}} = \frac{\sum_{i=1}^{k} P_i . N_{be,i}}{\sum_{i=1}^{k} N_{be,i}} \quad (1)$$

where, N_s is the total number of wet-deck slams, N_{be} is the total number of bow entry events, $N_{S,i}$ is the number of wet-deck slams in cluster *i*, $N_{be,i}$ is the number of bow entry events in cluster *i*, and P_i is wet-deck slamming probability of cluster i.

The number of bow entries for each cluster, $N_{be,i}$, in a seaway is a parameter that can be obtained through an ML model by considering a set of known features. What is difficult to estimate is the probability of wet-deck slamming for each cluster (P_i). The ideal situation is to find clusters in which P_i is the element of an ideal set, for example, clusters with $P_i \in \{1, 0\}$. This could be the case to a certain degree if features such as bow vertical acceleration and whipping frequency were used in an ML model such as the unsupervised model discussed in Section 4.1. As mentioned, such model is useful for automated data analyses for a monitoring system as described in Figure 7.

For simulation purposes, relative motions are considered as features to determine the rate of wet-deck slamming occurrences, in which P_i should be estimated for each cluster.

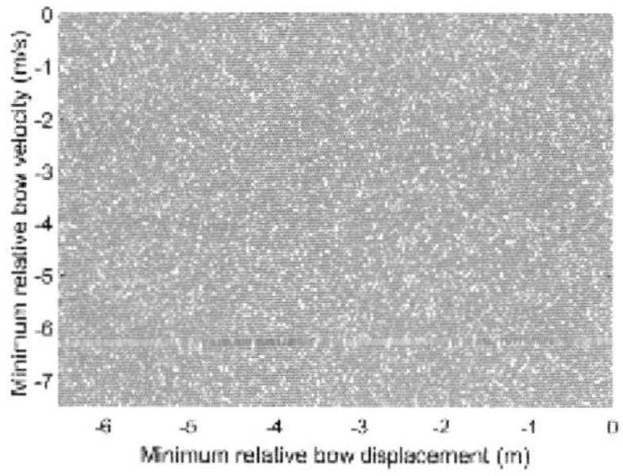

(a) Unclassified domain

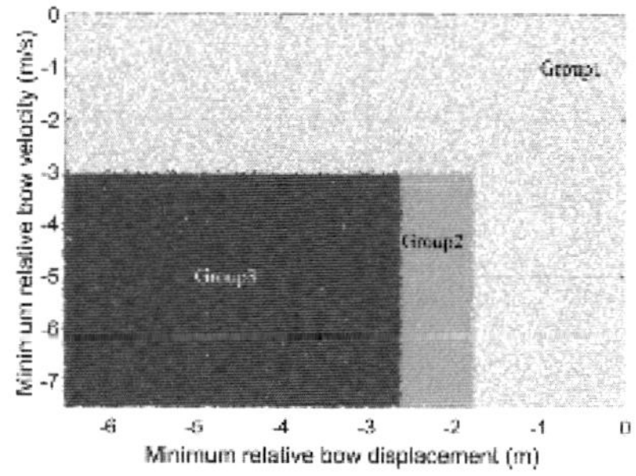

(c) Classification based on a coarse decision tree

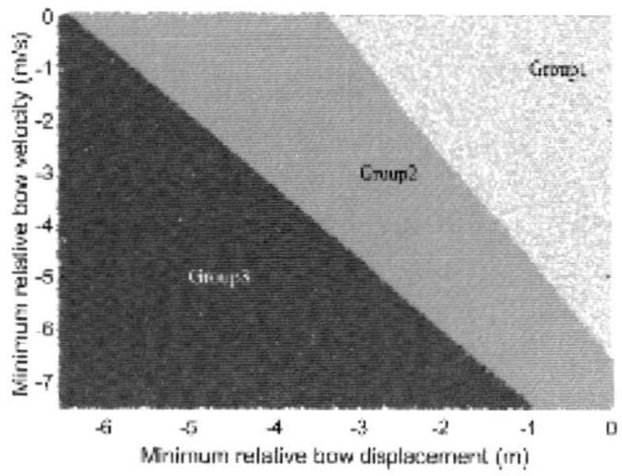

(b) Classification based on a linear support vector machine (SVM) model

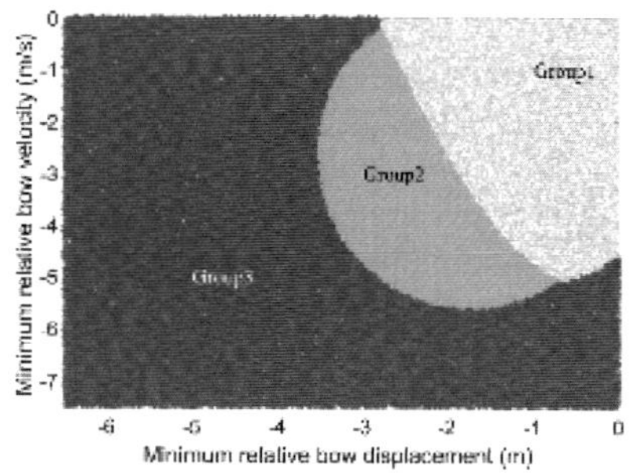

(d) Classification based on a Gaussian naïve Bayes model

Figure 17 An overview of bow entry classification using three supervised learning models by incorporating data shown in Figure 16 as training examples.

Table 6 Confusion matrix and accuracy of trained models

	Model 1	Model 2	Model 3
Confusion matrix (observations)	**Model : Linear SVM** True class 1: 1292, 80, – True class 2: 83, 615, 42 True class 3: 14, 139, 113 Predicted class 1, 2, 3	**Model :Coarse Tree** True class 1: 1327, 45, – True class 2: 250, 397, 93 True class 3: 29, 113, 124 Predicted class 1, 2, 3	**Model: Gaussian Naive Bayes** True class 1: 1307, 62, 3 True class 2: 113, 573, 54 True class 3: 16, 134, 116 Predicted class 1, 2, 3
Accuracy	84.9 %	77.7%	83.9%

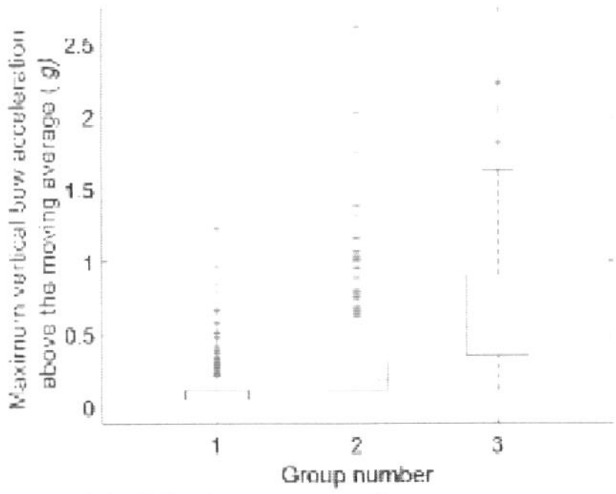

(a) Model 1 groups: linear support vector machine (SVM)

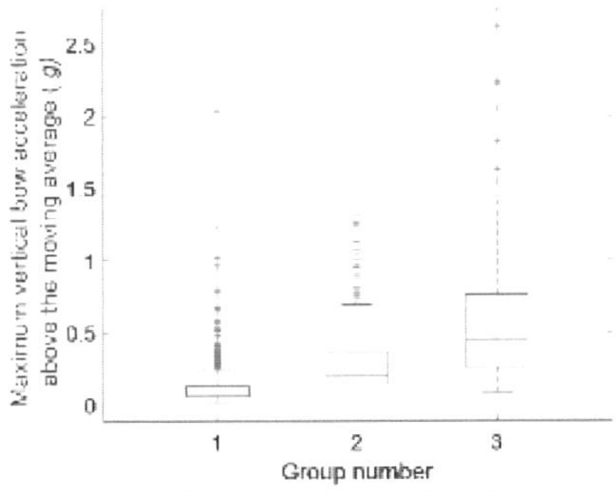

(b) Model 2 groups: coarse decision tree

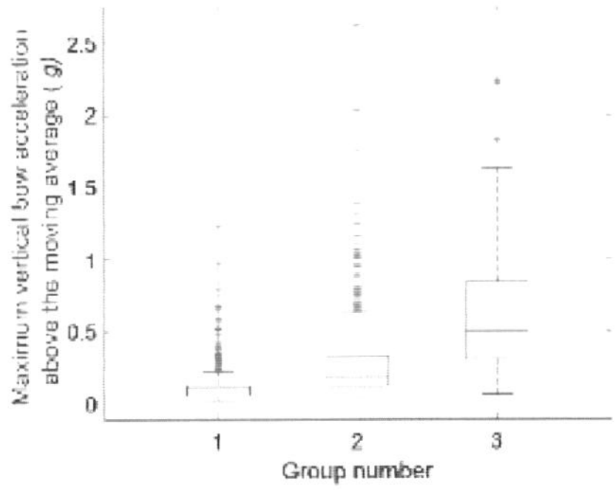

(c) Model3 groups: Gaussian Naive Bayes model

Figure 18 Box plot presentations of maximum vertical bow acceleration above the moving mean (*var2*) based on the outputs of three different classifiers trained by incorporating data shown in Figure 16 as training examples and general patterns described in Figure 17.

It is worth noting that, in the case of bottom slamming of monohulls, the assumption is that relative displacement and velocity are two features that describe "slamming" and "non-slamming" zones [16, 17, 28], which means there exist two clusters with $P_1 \cong 1$ for slamming events and $P_2 \cong 0$ for non-slamming events. However, this was not the case in Figure 18 (a-c) and a broader approach should be developed.

One approach to estimate $\boldsymbol{P_i}$ for clusters could be based on empirical cumulative distribution probability of a third-party feature which can describe the severity of slamming, such as maximum vertical bow acceleration during the bow entry or maximum vertical bow acceleration above the moving mean. Figure 19 shows cumulative probability plots of maximum vertical bow acceleration above the moving mean in three different groups obtained from the linear SVM classification model, which compares the cumulative probability at an arbitrary reference value of 0.5 $\boldsymbol{g}$. The cumulative probability of bow entry with a figure above this can be calculated as an estimate for $\boldsymbol{P_i}$ for each cluster, which are approximately, 0.01, 0.13 and 0.53 for groups 1, 2 and 3 respectively. Table 7 shows the estimations of $\boldsymbol{P_i}$ for each cluster for the three models trained.

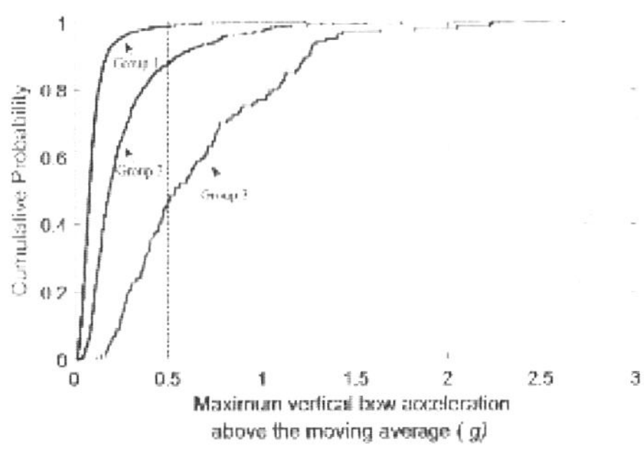

Figure 19 cumulative probability plots of max vertical bow acceleration above the moving mean in three different groups obtained from Model 1 classification algorithm (linear support vector machine (SVM)).

As can be seen in Table 7, wet-deck slamming events are less likely to occur in groups 1 and 2 compared to group 3 for each model. Although the choice of model does affect the probability of slamming for each group (i.e. $\boldsymbol{P_i}$), the number of bow entries in each group (i.e. $\boldsymbol{N_{be,i}}$) predicted by each model will change proportionally, and therefore $\sum_{i=1}^{k} \boldsymbol{P_i} \cdot \boldsymbol{N_{be,i}}$ is expected to be similar. Consequently, given a certain threshold (e.g. 0.5 $\boldsymbol{g}$ or 1 $\boldsymbol{g}$) the choice of clustering should not significantly change the probability of slamming $\boldsymbol{P}(\textbf{slam})$, which is equivalent to $\frac{\sum_{i=1}^{k} P_i \cdot N_{be,i}}{\sum_{i=1}^{k} N_{be,i}}$ as defined in Equation 1.

It is worth noting that the number and type of clusters can be important for optimal decision making, for instance to decide how often and to what degree the ship speed is required to be reduced in order to lower the probability of experiencing slamming events above a certain threshold in terms of slam-induced acceleration. However, recommending an optimal ship speed requires a much more complex matrix as other operational factors will come into effect.

Table 7 Empirical estimations of wet deck slamming probability in each group for different trained models with two thresholds of 0.5 ***g*** and 1 ***g*** for Hull 061 in head seas

	Group 1	Group 2	Group 3	Group 1	Group 2	Group 3
threshold	0.5 *g*				1 *g*	
Model1	0.013	0.130	0.530	0.001	0.032	0.250
Model2	0.016	0.150	0.440	0.002	0.052	0.190
Model3	0.013	0.140	0.490	0.001	0.037	0.220

5. CONCLUSIONS

A remote hull monitoring system was developed for a 111 m catamaran with the objective of 4G connectivity to a cloud-based platform in which machine learning/deep learning models could be deployed, enabling near real time analyses, classification and visualisation of data. The monitoring system is capable of measuring strain across multiple locations, bow and passenger deck accelerations, motions (i.e. heave, pitch and roll) and motions at the bow area relative to the water surface.

Using sea trials data available from a similar vessel and with the consideration of both unsupervised and supervised learning algorithms, centre bow entry events were classified into different groups and the likelihood of wet-deck slamming in head seas was estimated in each group for slamming events above two slam-induced bow acceleration thresholds of 0.5 ***g*** and 1 ***g***. By collecting a large amount of data of bow entry and wet-deck slamming events in various operational conditions, the models and estimates could be improved, and the probability of wet-deck slamming events in each cluster can be better evaluated.

More investigations are recommended to explore the application of learning models and recommendation systems for slamming, seakeeping, passenger comfort and the development of smart and connected hull monitoring systems.

6. ACKNOWLEDGEMENTS

This work was undertaken in collaboration between the University of Tasmania, Revolution Design, Incat Tasmania, University of New South Wales Sydney, Italian National Research Council and University College London through the support of the Australian Research Council Linkage Grant number LP170100555. The work of Mr. Pete Woodward in the development of the hull monitoring system is also gratefully acknowledged.

7. REFERENCES

[1] Lavroff J, Davis MR, Holloway DS, Thomas G. Wave slamming loads on wave-piercer catamarans operating at high-speed determined by hydro-elastic segmented model experiments. Marine structures. 2013;33:120-42.

[2] Lavroff J, Davis MR. Slamming kinematics, impulse and energy transfer for wave-piercing catamarans. Journal of Ship Research. 2015;59:145-61.

[3] Shabani B, Lavroff J, Davis MR, Holloway DS, Thomas GA. Slam loads and kinematics of wave-piercing catamarans during bow entry events in head seas. Journal of Ship Research. 2018;62:134-55.

[4] Shabani B, Lavroff J, Davis MR, Holloway DS, Thomas GA. Slam loads and pressures acting on high-speed wave-piercing catamarans in regular waves. Marine Structures. 2019;66:136-53.

[5] Shabani B, Lavroff J, Holloway DS, Davis MR, Thomas GA. The effect of centre bow and wet-deck geometry on wet-deck slamming loads and vertical bending moments of wave-piercing catamarans. Ocean Engineering. 2018;169:401-17.

[6] Thomas G, Winkler S, Davis M, Holloway D, Matsubara S, Lavroff J, et al. Slam events of high-speed catamarans in irregular waves. Journal of Marine Science and Technology. 2011;16:8-21.

[7] Amin W, Davis MR, Thomas GA, Holloway DS. Analysis of wave slam induced hull vibrations using continuous wavelet transforms. Ocean Engineering. 2012;58:154-66.

[8] McVicar J, Lavroff J, Davis MR, Thomas G. Fluid–structure interaction simulation of slam-induced bending in large high-speed wave-piercing catamarans. Journal of Fluids and Structures. 2018;82:35-58.

[9] Lavroff J, Davis MR, Holloway DS, Thomas GA, McVicar JJ. Wave impact loads on wave-piercing catamarans. Ocean Engineering. 2017;131:263-71.

[10] Davis MR, French BJ, Thomas GA. Wave slam on wave piercing catamarans in random head seas. Ocean Engineering. 2017;135:84-97.

[11] AlaviMehr J, Lavroff J, Davis MR, Holloway DS, Thomas GA. An experimental investigation on slamming kinematics, impulse and energy transfer for high-speed catamarans equipped with Ride Control Systems. Ocean Engineering. 2019;178:410-22.

[12] French BJ, Thomas GA, Davis MR. Slam occurrences and loads of a high-speed wave piercer catamaran in irregular seas. Institution of Mechanical Engineers Proceedings Part M: Journal of Engineering for the Maritime Environment. 2013:1-13.
[13] Shabani B, Holloway D, Lavroff J, Davis M, Thomas G. Systematic model tests on centre bow design for motion and slamming load alleviation in high speed catamarans. 14th international conference on fast sea transportation, 2017. p. 136-43.
[14] Shabani B, Lavroff J, Holloway DS, Davis MR, Thomas GA. Wet-deck slamming loads and pressures acting on wave piercing catamarans. International Shipbuilding Progress. 2019;66:201-31.
[15] Shabani B, Lavroff J, Holloway DS, Davis MR, Thomas GA. The influence of the centre bow and wet-deck geometry on motions of wave-piercing catamarans. Proceedings of the Institution of Mechanical Engineers, Part M: Journal of Engineering for the Maritime Environment. 2019;233:474-87.
[16] Ochi MK, Motter LE. Prediction of slamming characteristics and hull responses for ship design. 1973.
[17] Dessi D, Ciappi E. Slamming clustering on fast ships: From impact dynamics to global response analysis. Ocean Engineering. 2013;62:110-22.
[18] Magoga T, Aksus S, Cannon S, Ojeda R, Thomas G. Identification of slam events experienced by a high-speed craft. Ocean Engineering. 2017;140:309-21.
[19] Aghabozorgi S, Shirkhorshidi AS, Wah TY. Time-series clustering–A decade review. Information Systems. 2015;53:16-38.
[20] Jain AK. Data clustering: 50 years beyond K-means. Pattern recognition letters. 2010;31:651-66.
[21] Kubat M. An introduction to machine learning: Springer; 2017.
[22] Mahdavinejad MS, Rezvan M, Barekatain M, Adibi P, Barnaghi P, Sheth AP. Machine learning for Internet of Things data analysis: A survey. Digital Communications and Networks. 2018;4:161-75.
[23] Witten IH, Frank E, Hall MA, Pal CJ. Data Mining: Practical machine learning tools and techniques: Morgan Kaufmann; 2016.
[24] Mitra M, Gopalakrishnan S. Guided wave based structural health monitoring: A review. Smart Materials and Structures. 2016;25:053001.
[25] Jacobi G, Thomas G, Davis MR, Davidson G. An insight into the slamming behaviour of large high-speed catamarans through full-scale measurements. Journal of Marine Science and Technology. 2014;19:15-32.
[26] Jacobi G, Thomas G, Davis M, Holloway D, Davidson G, Roberts T. Full-scale motions of a large high-speed catamaran: The influence of wave environment, speed and ride control system. International Journal of Maritime Engineering. 2012;154:A143-A55.
[27] Davis M, French B, Thomas G. Wave slam on wave piercing catamarans in random head seas. Ocean Engineering. 2017;135:84-97.
[28] Dessi D. Whipping-based criterion for the identification of slamming events. International Journal of Naval Architecture and Ocean Engineering. 2014;6:1082-95.
[29] Amin W, Davis M, Thomas G, Holloway D. Analysis of wave slam induced hull vibrations using continuous wavelet transforms. Ocean Engineering. 2013;58:154-66.
[30] Jain AK, Murty MN, Flynn PJ. Data clustering: a review. ACM computing surveys (CSUR). 1999;31:264-323.
[31] Berkhin P. A survey of clustering data mining techniques. Grouping multidimensional data: Springer; 2006. p. 25-71.
[32] Jain AK, Dubes RC. Algorithms for clustering data. Englewood Cliffs: Prentice Hall, 1988. 1988.
[33] Singh A, Yadav A, Rana A. K-means with Three different Distance Metrics. International Journal of Computer Applications. 2013;67.
[34] Khan SS, Ahmad A. Cluster center initialization algorithm for K-means clustering. Pattern recognition letters. 2004;25:1293-302.
[35] Likas A, Vlassis N, Verbeek JJ. The global k-means clustering algorithm. Pattern recognition. 2003;36:451-61.
[36] Bezdek JC. Pattern recognition with fuzzy objective function algorithms: Springer Science & Business Media; 2013.
[37] Dunn JC. A fuzzy relative of the ISODATA process and its use in detecting compact well-separated clusters. 1973.
[38] Lloyd S. Least squares quantization in PCM. IEEE transactions on information theory. 1982;28:129-37.
[39] Arthur D, Vassilvitskii S. k-means++: The advantages of careful seeding. Proceedings of the eighteenth annual ACM-SIAM symposium on Discrete algorithms: Society for Industrial and Applied Mathematics; 2007. p. 1027-35.

8. AUTHORS BIOGRAPHY

Babak Shabani is postdoctoral researcher at the School of Engineering at the University of Tasmania, Australia. He is responsible for developing cloud-based machine learning systems specific to remote monitoring of high-speed catamarans. His previous experience includes "slam loads and kinematics of wave-piercing catamarans during bow entry events in head seas", the winner of Vice Admiral E. L. Cochrane Award (2019) from SNAME.

Dr **Jason Lavroff** holds the current position of senior lecturer at the University of Tasmania. He is responsible for teaching in dynamics, vibrations and design together with research on motions and loads of high-speed catamaran vessels. His previous experience includes extensive industry collaboration with Incat Tasmania on research involving ship motions and loads both at model scale as well as full-scale with a particular focus on slamming using both theoretical and experimental techniques. More recently this work has been extended by machine learning and classification at full-scale that is of particular relevance to the current paper.

Damien Holloway holds the position of Associate Head Learning and Teaching in the School of Engineering, University of Tasmania and has responsibility for coordinating all of the university's engineering undergraduate and honours programs. He conducts research in the areas of structural health monitoring, structural dynamics, buckling and acoustics, and he teaches structural mechanics, vibrations and finite element analysis. He has published around 70 peer-reviewed journal articles, supervised 12 PhDs to completion, and has been a core member of the University of Tasmania's high-speed catamaran research group collaboration with shipbuilder Incat for around 25 years.

Spiridon Penev is an Associate Professor at the Department of Statistics, UNSW Sydney, Australia. He is teaching Statistics courses at both undergraduate and postgraduate level and is supervising honours and postgraduate students. His research is in statistical methodology in general and in nonparametric statistics with emphasis on wavelet methods, in particular.

Daniele Dessi is Senior Research Scientist at the Institute of Marine Engineering at the National Research Council of Italy where he is responsible for the Laboratory of Dynamics and Diagnostics of Structures. He holds a Master of aeronautics and a PhD in Aerospace Engineering from the University of Rome "La Sapienza". His previous experience includes fluid-structure interaction, hydroelasticity, structural health monitoring, scaled model testing and energy harvesting from vibrations and since 2006 he has been a member of the International Ship and Offshore Structures Congress (ISSC).

Professor **Giles Thomas** is currently the BMT Chair of Maritime Engineering at University College London. A naval architect, he has over 28 years of research and development experience in the fields of hydrodynamics and fluid-structure interaction working with a large range of industry partners. He is currently the Technical Lead for the EU project SEDNA: Safe maritime operations under extreme conditions: the Arctic case.

FUTURE WORKFORCE AND TRAINING IN THE FOURTH MARITIME REVOLUTION

R. Westgarth and **J. Rigby**, BMT, UK

SUMMARY

The fourth Maritime Industrial Revolution is driven by emerging and disruptive technologies. The key drivers for the coming decade are expected to be digitalization and decarbonisation leading to the emerging focus on connected smart ships, autonomous vessels and green shipping technologies. Emergent technologies offer opportunities for the maritime industry to become safer and more efficient while reducing its environmental footprint. This will transform the industry, changing its more traditional trading ecosystem. This will create a new range of opportunities for the Maritime Industry but will also create significant training and workforce challenges.

1. INTRODUCTION

The Fourth Maritime Industrial Revolution is driven by emerging and disruptive technologies. The key drivers for the coming decade are expected to be digitalization and decarbonisation leading to the emerging focus on connected smart ships, autonomous vessels and green shipping technologies. Emergent technologies offer opportunities for the maritime industry to become safer and more efficient while reducing its environmental footprint. This will transform the industry, changing its more traditional trading ecosystem to a symbiotic one linking familiar innovative concepts such as the Digital Economy, Future Cities and new trading systems to create Shipping 4.0 or the Fourth Maritime Revolution.

The future workforce will need to adapt to this changing work environment, with changing working attitudes and new skills brought about by new technological and demographic changes. New roles and jobs are expected to be created as more traditional roles are scaled down with the increased use of technologies like Artificial Intelligence and Robotic systems. Human users need new skills achieved through new forms of training to keep up with the high speed of change. The systems likewise need to adapt; becoming more human focused in their design and operation. The reshaping of the workforce, training and technology are key to unlocking the productive potential of the Fourth Industrial Revolution.

2. THE FOURTH MARITIME INDUSTRIAL REVOLUTION AND THE FUTURE WORKFORCE

The Fourth Industrial Revolution will generate a great deal of change that the maritime industry needs to prepare for. It is not just the traditional manufacturing sectors of the industry that will be affected; the service sector faces significant disruption as global demand in consumer and commercial market drives the use of Artificial Intelligence, Machine Learning and Automation (Figure 1). Businesses, regardless of their role, need to anticipate change across four areas: the industry, the workforce, the businesses and the labour landscape.

Figure 1: Autonomous Technology is poised to reshape the maritime sector, like this unmanned containership concept.

As additive manufacturing increases in popularity and uptake we could see the industry face change as the nature and focus of trade shifts along emerging geopolitical schisms (Figure 2). The change, enabled by emerging technology would allow for increased collaboration and connectivity across continents. High technology companies similar to Amazon or Google are expected to enter the maritime industry (Riviera, 2019) and upset the status quo, revolutionising the otherwise traditionally structured supply chains and increasing the adoption of user-friendly technologies. It is anticipated this will drive a proliferation of low cost technology and upend the present conditions of the Maritime Industry.

The industry's workforce is changing also; a skills shortage across the industry, caused by shorter seafaring careers and a greater reluctance of people to serve on vessels (The Maritime Executive, 2019), threatens to worsen and the skillsets required by workers is changing. The specific challenges faced by the maritime industry are compounded by socio-demographic changes across the wider workforce. The workforce is aging, multigenerational, more international and diverse (World Economic Forum, 2018). Record numbers of workers over 50 are retiring and the workforce is losing valued skills. As careers are becoming shorter they are also becoming more varied as individuals gain and carry skills across multiple sectors.

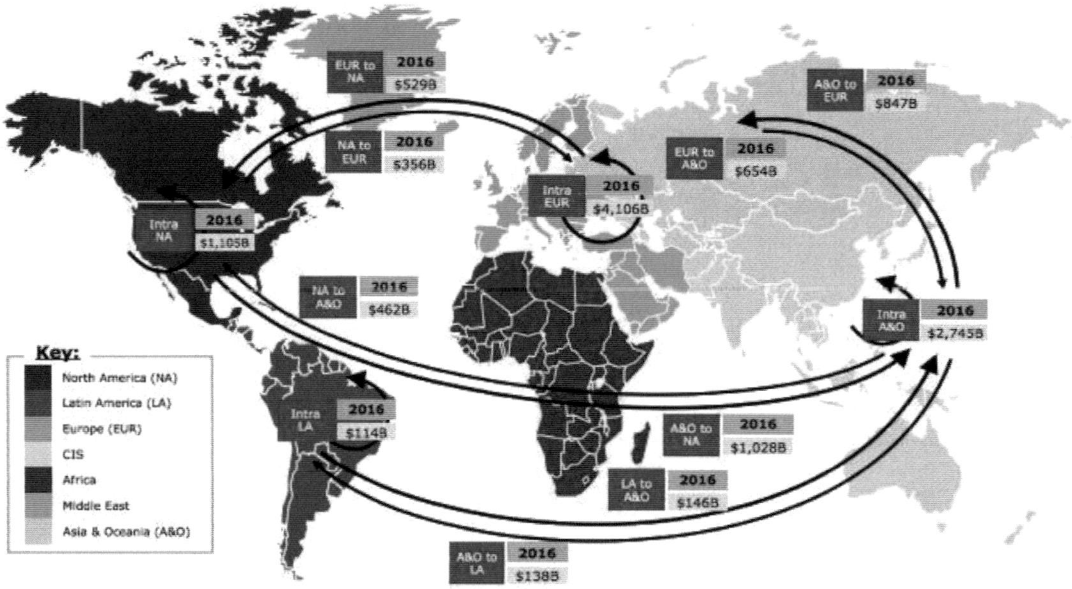

Source: WTO world region export analysis, October 2017; Bain & Company analysis

Figure 2: Global Trade Flows are Large and Increasing (WTO, 2019).

Increasingly flexible conditions are expected to be demanded with workers looking to improve their work / life balance. Digitalisation and virtualisation of the work place has the possibility to enable this across all areas of business, making the maritime industry more attractive in the world skills market.

As the industry, workforce and businesses change so too does the labour landscape and market. Whilst it is impossible to predict what it will comprise of in 20 years there are indications of its direction of travel. Large numbers of existing jobs could become redundant whilst the skills required for the others will change dramatically. WEF Future of Jobs Report 2018 (World Economic Forum, 2018) suggests that 50% of the workforce will need reskilling by 2022 as the labour landscape changes.

The WEF report also predicts that 65% of children attending school today will go into jobs that don't yet exist. Bearing this in mind, there presents a clear requirement for updating the skills of the present workforce. As the rate of technological innovation and introduction increases so too will the need for updating the workforce's skills. The traditional training model – an early career formal education or training program – is becoming redundant. A lifelong approach to learning needs to be adopted with individuals continually staying abreast of technological developments. The training subjects shouldn't be traditional either; WEF identifies new technological competency and critical human skills such as creativity, originality and complex problem solving as the core skills required in the future labour landscape.

The main challenge, therefore, of the revolution is capability management. Whilst technological development drives what is produced, the workforce's technical and human capabilities dictate how much can be produced. With capability providing the limiting factor in productivity, the workforce needs to adapt through the early identification of future skill requirements and promptly training its members to utilise new technologies effectively.

3. CHALLENGES

So how do we see the challenges for this future workforce. The Global Marine Technology Trends Report produced by QinetiQ, Lloyds Register and University of Southampton (2015) identified a range of 18 technologies in 3 areas, commercial shipping, defence and ocean science. When considering the future workforce, we can see four trends that we should prepare for:

1. Digitalisation – the growth of data, analytics and AI are expected to change the skills required for traditional operations. This change is set to accelerate as the impact of the proliferation of technology at low cost driven by the consumer markets.

2. Connected Technologies – systems and equipment will be more dynamic and increasingly interconnected as communications and networking technologies accelerate (e.g. 5G Communications and the Internet of Things).

3. Emission control, reduction and ultimately alternative energy solutions – are expected to come to the market. As the sustainability and emissions targets become more challenging.

Figure 3: Increasing regulation is forcing ship operators to reduce their emissions using new fuels and adding scrubber systems to their exhausts.

4. Demographics, skills and learning – in particular the impact of the Changing nature of workforce and use of robotics. The Maritime sector as a whole will need to adapt as the impact of global demographic change, the rise of middle classes and an ageing population are felt.

The development of smart ships, unmanned ships (autonomous or remote controlled), port automation and digitalisation in the service elements of the sector will force change to the ability to recruit and retain seafarers, increasing competition for scarce talent with other emerging digitised businesses in all sectors, coping with rapid technology change will force change to existing structures, people, career paths

In addition, the ageing population, continuing shortages in engineers and scientists will impact upon the traditional nature and culture of the Sector will give rise to mismatches in a broader "18 to 80" workforce, the multigenerational nature could highlight disparities in career expectations, technology abilities and at a fundamental level, behaviours and values.

This future is full of uncertainties and challenges which we must overcome, requiring future maritime leaders in government, the military and industry to develop new and effective leadership capabilities, leading not just the people, but also the machine / human partnerships that will underpin defence capability.

BMT is conscious that we need to be proactive in shaping our future. As a company we cannot wait for events to happen, we need to continue to develop ourselves both as a business and as a workforce so that we are prepared for this future and, just as importantly, we are in a position to help our customers in the Enterprise, prepare for this future. A key element of this will be the future of leadership across the sector. The challenge will be in developing capable and effective leadership throughout the enterprise in these changing times.

We need to recognise that the enterprise now extends from service sectors, shipping and transportation companies into major suppliers and their supply chain and down into Small and Medium Enterprises. The reliance on cutting edge technology will require new concepts to be developed in the business operations. Our workforces will be increasingly interconnected. The competition for talent will require us to reflect on the traditional organisational and commercial models and many will need to change if we are to successfully continue to deliver effective maritime outcomes.

Given a changing workforce with far more complex and diverse notions of identity, loyalty and expectations of meaningful and purposeful work, leaders will need to be more flexible adapting approaches appropriate for the differing audiences they will lead.

The modern workforce wants and needs, are a far more modular experience. Gen Z acknowledge that careers change rapidly (Bloomberg, 2019), and many of today's careers will not exist throughout their lives. They are often looking for modular careers, where they acquire skills that will allow them to change roles regularly. Equally, many skilled people are now finding jobs they have spent their lives in are becoming obsolete and considering retraining into new environments. This transformation of the workforce and the workplace will place heavy demands on future leaders, they will need to be capable of leading through continuous and sometimes turbulent change, constantly learning and evolving as individuals are adapting to emergent challenges.

This is a massive societal challenge, that the enterprise is currently ill-suited to dealing with. The traditional hierarchies where pay and rank is related to experience must urgently change. New employment models must be developed, and job roles created, which allow people to transition in and out of the military throughout their careers whilst progressing and learning new skills.

In summary, fundamental to dealing with the change brought about by the fourth maritime industrial revolution will be re-imagining the way we recruit, train and develop our people. To do this we will need to consider:

- How does the maritime enterprise respond to these challenges and opportunities to create a workforce of the future?
- How does it create a public and private sector environment where new processes, leadership and management create new ways of thinking, and allow new ideas and technologies to be exploited?
- How does it become more agile and data driven?
- How does it attract and support the right people?

4. SHAPING THE RELATIONSHIP BETWEEN PEOPLE AND TECH

BMT are actively addressing many of these issues, engaging with leading organisations around the world to better understand the nature and scale of disruption identified above, its impact upon our business and our clients and more importantly the needs of our people.
We are looking at how this future relationship between people and technology will reshape the industries and clients we work with, developing insights and thought leadership on related topics.

We are addressing this by extending our strategic planning horizons to enable a long-term transformational approach to workforce capability planning that 'reads' strategic trends to anticipate their impact on the 'future of work', and responsively adapt recruitment, selection, training, deployment and retention. For this we need to be pragmatic, recognising that planning should not be based on a single, 'perfect' future, but focused on delivering resilience against a range of possible future scenarios.

We have developed a series of strategic campaigns looking at the transformational nature of technologies such as digitisation, autonomous systems, AI and Machine Learning, cyber etc to both understand, identify the potential impact, consider our own skills needs and just as importantly the likely needs of our customers.

Other activities include collaboration with AI and robots. It is understood that the move to autonomous ship navigation if implemented will not occur overnight. Instead, for a period of years and even decades manned vessels will need to continue to operate alongside and in partnership with their autonomous counterparts. In order to facilitate this BMT have teamed up with L3 ASV to explore and understand how autonomous navigation systems can be improved for greater predictability and safety in confined busy waterways. Using BMT's REMBRANDT simulator to train and validate L3 ASVs vision algorithms allows the user and regulators to gain trust in the system before it is realised into the open water.

Figure 4: Simulated environments like BMT's REMBRANDT software will be required to test and validate autonomous navigation systems before they can be implemented.

BMT is also very passionate about the importance of human-centred design in new technological environments. As the systems become more and more complex, the analytical and physical demands on the human user also greatly increase. Our Human Factors team are at hand to support a wide range of projects and ensure the user is considered from the initial design. New technologies such as Virtual and Augmented Reality environments can be used to test the human user interfaces providing customers and users with rapid feedback on the implications of design choices. Enabling us to ensure and effective integration of technology and people, embedded in an ethical design approach.

Figure 5: Virtual and Augmented Reality environments can be used to test the human user interfaces with new technology concepts and products.

5. HUMAN AUTONOMY TEAMING AND TRUST

The evolution of work facilitated by technology development is not new. We are moving from an industrial age, which at its inception led to massive social disruption, to an automation age which promises the same. However, there's an opportunity to consider some of what history has to teach us. The future of work is likely to be one where workforce autonomy and intelligent systems are common place.

The utopian view of the industrial age was that "technology was supposed to free people for more productive uses of their time". In a corporate setting this goal has generally been subverted as companies invested heavily in new tech, but neglected integration and usability, integrating tools with one another (though not always effectively) but not with the way people work. As a result, corporate technology often creates frustration and lowers productivity – there is a risk that workplace automation simply reinforces this trend.

However, there are opportunities to generate new jobs, providing opportunities for those displaced and exciting opportunities for future generations who are already far more technically capable than their parents.

Here in BMT we have a rare breath of capability that already links technology development and behavioural science offering the option of extending our reach into alternative markets. If we have the foresight and agility to adapt to changing demands as new opportunities emerge. The advent of advanced manufacturing technologies supported by intelligent systems represent opportunities to reduce the costs of maritime solutions, enabling greater probability of exports and enabling the UK to compete with low cost labour markets through the exploitation of hi tech affordable solutions.

New solutions will require new methods for design and manufacture such as advanced simulation. The use of non-deterministic capabilities, such as machine learning will require the development of alternative engineering paradigms, new skills and different roles within the companies creating future concepts for learning and development of both people and the system.

This new technological era, combined with new demographic generations and disruptive business models offers BMT a real opportunity in the medium and long term to develop as our customer's key advisor, capable of integrating technological, industrial and human capabilities in new ways, working with new innovative businesses to deliver business advantage in a turbulent world.

As the adoption of emerging technology accelerates, driven by large technology corporations, an emerging challenge is building the trust in the systems by the people and society that they impact upon. But how do you achieve it and how much trust is enough to warrant acceptable deployment? This people element is often overlooked in our race for technology.

Autonomy is a key technology enabling driverless cars, new business models, new methods of transport and requiring high levels of interaction between not just between the people but also between the infrastructure of future cities. We need to recognise that across our societies this will have a polarising impact on people's views. To overcome this, we need to understand the role of trust in driving these perspectives and learn how to demonstrate to our citizens that these systems are safe.

The technology is seen by many as the big opportunity, but it is also seen as an existential threat to societies, communities and jobs. The focus is to increase productivity, lower operating costs, drive up economic value. However, have we given enough thought to the impact upon our people? We do know that not all roles will be at risk of automation. Such as jobs that are low-skilled, but require high dexterity. It's unlikely that entire roles will be automated. The consensus view is that automation will affect every job category by at least 25%.

6. CONCLUSIONS

Whilst the future is uncertain there are inferences as to what it may entail: greater use of technology and a greater emphasis on human and technical skills of the workforce. Individuals and businesses need to adapt by adopting a lifelong approach to learning and improving their adaptability to changing circumstances. Future technologies have immense potential but the narrative between people and technology needs to change to a more holistic view. The combination of user friendly technology and a highly capable workforce will unlock the productive potential of the Fourth Industrial Revolution in ways just focusing on technological advancement could not.

As the sector embraces the changes described above, the maritime sector becomes even more relevant particularly from a global prosperity and environmental compliance perspective, as highlighted by the OECD Report "The Ocean Economy in 2030" (OECD, 2016). Indeed, the future of the sector is looking particularly buoyant and this is evidenced by the ever-increasing adoption of emerging technology in the sector. The demands placed on our oceans and waterways are and will continue to increase, with population demands (particularly in our coastal cities) driving transport loads. The scarcity of natural resources will give rise to new areas of activity such as seabed mining and exploration. As an industry, we will need to take a collaborative and coordinated approach in order to achieve this safely and without over burdening the sector or resources available. There are two overriding aspects that the sector needs to overcome to facilitate this; they are:

- To exploit technologies throughout the supply chain, such that it both enables and supports business and generates improved safety and productivity delivering significantly increased business performance and enabling innovation within the sector. As the individual technologies mature, systems will undertake an increasing number of functions. We will need to address the higher-level areas such as operational planning and management and control together with the integration of expanding data and information capabilities to provide cost effective solutions.
- These changes will need to be carefully managed and appropriate regulatory and governance frameworks put in the place to support their development and the people, a key factor will be the ability of regulators and similar organisations to "match the pace" of technological change, this will include:
    - Understanding the interactions between people and intelligent systems, not just from a maintenance or control perspective, but also wider socio-economic perspectives such as the impact on skills and jobs.
    - Recognise and address the challenges in transitioning the industry over the coming decades from today's solutions to this future autonomous and intelligent nirvana

As we are seeing in other sectors, new business models will lead to the potential disruption of traditional organisational models, the sector has already recognised the declining availability of personnel willing to take jobs, this coupled with ongoing demographic change is set to continue and even accelerate, the adoption of technology will potentially be the only way to keep trade and business functioning.

Looking even further ahead at longer term technological impacts, it is predicted we will see the fundamental change of the traditional "factory to consumer" distribution models. The advent of additive manufacturing, whilst at best a nascent technology today, will enable production at the "point of use", changing the balance between low wage / low cost manufacturing centres to localised manufacturing at the point of need, often enabling a high degree of "bespoking" of individual items.

Thus, the maritime industry will undergo a profound and deep change as technologies which enable the Fourth Industrial Revolution mature and businesses take competitive opportunities for their exploitation. In terms of jobs, we see a significant re-shaping of the workforce, moving away from traditional manual roles and activities into more supervisory roles, coupled with increasing technological capabilities, this will result in the need to generally re-skill elements of the workforce.

7. ACKNOWLEDGEMENTS

The authors would like to acknowledge all those who supported the future workforce research conducted at BMT and the production of this paper.

8. REFERENCES

1. Bloomberg. (2019). Millennials, Gen Z Workers Change Jobs Rapidly, Says CareerBuilder CEO. [online] Available at: https://www.bloomberg.com/news/videos/2019-10-04/millennials-gen-z-workders-change-jobs-rapidly-says-careerbuilder-ceo-video [Accessed 11 Sep. 2019].

2. OECD (2016). The Ocean Economy in 2030.

3. Riviera. (2019). Google enters the autonomous shipping race 1. [online] Available at: https://www.rivieramm.com/news-content-hub/news-content-hub/google-enters-the-autonomous-shipping-race-1-53060 [Accessed 11 Sep. 2019].

4. The Maritime Executive. (2019). The Future of the Modern Seafarers? Profession. [online] Available at: www.maritime-executive.com/editorials/the-future-of-the-modern-seafarers-profession.

5. University of Southampton (2015). Global Marine Technology Trends. University of Southampton.

6. World Economic Forum. (2018). The Future of the Modern Seafarers? Profession. [online] Available at: https://www.weforum.org/reports/the-future-of-jobs-report-2018

9. AUTHORS BIOGRAPHY

Richard Westgarth Richard has over 40 years of experience in the defence and maritime sectors as an innovation strategist and business adviser. His wealth of experience in naval and defence domains builds on his earlier work exposure in research, innovation, change management and consultancy. His role as Head of Campaigns at BMT is to operate across all the Organisation's Partnerships, looking at medium to long term trends and developing BMT's cross-functional capabilities to match the customers future needs.

Jake Rigby holds the current position of Research and Development Lead at BMT. He is responsible for Research Portfolio Management, Academic Engagement and Horizon Scanning. His previous experience includes 4 years as a Naval Architect working on a range of ship designs and consultancy projects.

DIGITALISATION IN THE UK MARITIME SECTOR: A STAKEHOLDERS' PULSE CHECK

C. Gkerekos and **G. Theotokatos** and **L. M. Bujorianu** and **E. Boulougouris** and **D. Vassalos**,
Maritime Safety Research Centre, University of Strathclyde, Glasgow, UK

B. Carballedo and **S. McCluskey**,
HSSMI, Stratford, UK

SUMMARY

Several digitalisation technologies have recently emerged, reshaping organisations across various industries and sectors. Such technologies, commonly placed under the industry 4.0 umbrella term, include Internet of Things (IoT), cloud computing, blockchain, and Artificial Intelligence (AI). They aim to enable the data collection, exchange, processing, and automation to enable a system to make decisions autonomously. This paper presents IN 4.0 project, aiming to ease the adaptation and adoption of industry 4.0 practices and technologies in maritime sector companies in order to improve their competitiveness. This will be achieved through a four-pronged approach aiming to: analyse the current state-of-art as applied to other industries; define a protocol for the redefinition of workers' tasks; investigate cost-reduction methods for the implementation of 4.0 technologies; and counsel businesses for the transformation of maritime SMEs to the 4.0 era. Special focus will be given to the training required to get workers acquainted with new technologies and competitive in this changing labour market. Recent progress and advances of the IN 4.0 project along with future plans are presented and discussed, thus revealing the expected future transformation of the maritime industry.

1. INTRODUCTION

Ships are a crucial asset of the global goods transportation system, as 85% of merchandise is carried by sea (Clarkson PLC, 2018). However, the shipping industry has remained change-averse for a long time, only adopting new technologies when required by regulations or when a short-term financial benefit is evident.

Many other industries are rapidly changing, incorporating industry 4.0 practices and technologies stemming. Such technologies include Internet of Things (IoT), cloud computing, blockchain, and Artificial Intelligence (AI). The aim of industry 4.0 is to enable the data collection, exchange, processing, and automation to enable systems to make decisions autonomously.

These new technologies are trickling down to the shipping industry. Nonetheless, these changes often only affect the state-of-art, leaving the state-of-practice stationary. Furthermore, even when changes reach the state-of-practice (e.g. Inmarsat reports that the average surveyed ship-owning organisation plans to invest $2.5 million in IoT applications over the next 3 years (Inmarsat, 2018)), these are often only applied lackadaisically, leading to lacklustre results that, in turn, hinder future R&D investments.

This paper aims to present IN4.0 project, aiming to ease the adaptation and adoption of industry 4.0 practices and technologies in maritime sector companies in order to improve their competitiveness. More specifically, the project aims to promote the modernisation of the maritime sector through the implementation of actions that help to transform Small-Medium Enterprises (SMEs) to their 4.0 equivalents, identifying barriers that prevent the innovation of the maritime sector business model, improving the companies' production processes, transforming the work organisation systems, knowledge and commercialisation, as well as training workers of the maritime sector in new occupations and tasks.

2. BACKGROUND

The maritime sector occupies a central position in the growth of many countries' economy, being deeply rooted in the Atlantic Area countries and presenting a common problem: the urgency of getting contracts in a very competitive market, mainly caused by very low labour costs in European countries such as Romania, Turkey, Poland or Croatia, or at worldwide level as Asians, competing with the maritime companies of the Atlantic Area countries. To prevent a competitive advantage of the Atlantic Area maritime sector based on the reduction of wages, and considering the particularities of the small size of the sector companies, the joint solution to be applied must be based on technological improvements associated to the new industry 4.0 paradigm, supported by innovations such as optimised organisation or others, linked to training and commercialisation.

The above remarks solidify the need for a collaboration between industry and academia in order to sustain and improve the competiveness of the UK maritime industry. The innovative character of the project stems firstly from its own nature, because it addresses the issue of the innovation in the shipbuilding industry by adapting its business model towards the Industry 4.0. So far, the innovation has been oriented to achieve greater efficiency in the production process. However, the shipbuilding sector SMEs cannot implement the technologies required to adapt their business model without having a skilled workforce, an organisational structure adapted to the 4.0 industry processes, suitable marketing channels, a pre-selection of existing technologies and their degree of maturity, or the tools to reduce implementation costs in their workplaces. In this sense, IN 4.0 is an innovative project, aiming to boost the maritime sector, through the efficient use of human talent based on organisational innovation (creation of new functions / tasks of different jobs for an effective implementation of each one of the detected technologies), the design of a specific training model (that fulfils the training needs of the naval sector workforce in the context of factory 4.0: managers, middle managers, including young graduates) and the definition of new marketing methods. Furthermore, all the above will be implemented in and demonstrated through 10 SMEs selected among beneficiary countries of INTERREG Atlantic Area Programme.

3. IN4.0 FRAMEWORK

Project IN 4.0 aims to improve the competitiveness of naval sector companies through the adaptation of the traditional productive model towards the Industry 4.0, guaranteeing the sector's endurance in an increasingly demanding market where technological innovation is a key factor of strategic advantage.

To achieve this goal, the main barriers which impede the adaptation of the naval sector business model will be validated; determining the degree of maturation of existing technologies that could be implemented in naval SMEs; redefining work organisation systems; qualifying naval sector workers in order to move towards 4.0 model; designing innovative strategies to save costs in the implementation of technologies, detecting new marketing methods, and implementing the innovations designed under IN 4.0 in the selected SMEs of the participating regions in the project.

It is implemented through 4 technical work packages:

1. IN CONTEXT: starting point towards the adaptation to a new productive model 4.0 on the basis of the existing state of the art, detection of innovation applicable to the important 4.0 naval technologies relevant for the naval sector.
2. IN WORK AND IN TRAINING: Protocol for the redefinition of workers' tasks and training of sector staff.
3. IN COMMERCIALISATION AND COSTS: Reduction of the costs related to the implementation of technologies, and new methods of commercialisation.
4. IN ADAPTATION: Counselling for the transformation of naval companies into intelligent companies.

4. IN4.0 PROGRESS & ADVANCES

As part of the IN4.0 project, an event was held in Glasgow to determine the current state of the UK Maritime Sector and introduce the concept of Industry 4.0 to relevant stakeholders, namely technology providers, industry, funding bodies, and academics.

In order to achieve the goals of this event, attendants were asked to participate in a number of digital surveys to gauge their opinions on a number of topics. Furthermore, following the completion of each survey, each group of attendants was given 5 minutes to discuss the question in further detail with an aim of gathering key points and establishing commonalities across the industry. Some of the key topics discussed are presented below.

4.1 LEVEL OF INDUSTRY 4.0 ADOPTION

As shown in Figure 1, the survey showed distinct variance across the results with the majority of attendees at least scoping potential I4.0 implementation. It was noted that an additional choice of "Not ready yet" would have been more applicable with the current position of a number of businesses.

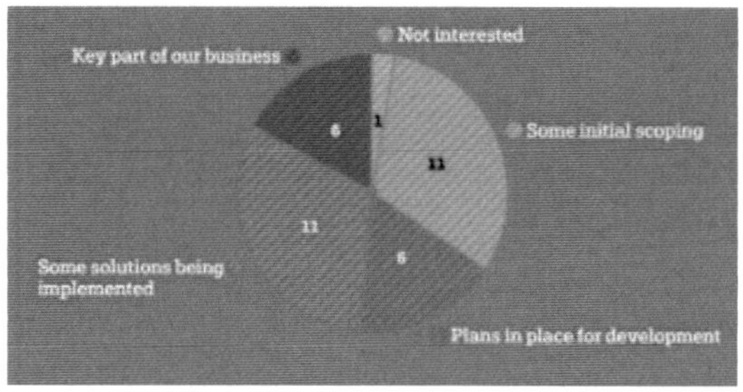

Figure 1: Survey results obtained on "Where do you feel your business is currently positioned on its I4.0 journey" question.

Some of the key points of the conversation that ensued are summarised below:

- Lack of strong leadership to drive forward innovative projects leading to unsuccessful implementation and disengagement.
- The industry is difficult to compare to others due to the inability to prototype, long lead times, and ship building operations commencing prior to complete vessel design.
- General consensus that the maritime sector is extremely risk adverse leading to a lack of willingness for developing new and innovative solutions.
- There is no clear technology strategy across many businesses in the sector.
- Concerns over an aging workforce who may not be technically adept, but possess extensive knowledge which could soon be lost to future generations in the industry.
- There is a lack of knowledge on the technologies within Industry 4.0 and the potential benefits to the business as well as strategy to develop solutions.
- Bureaucratic procedures can become a challenge when implementing a new technology. It was recognised that the level of technology adoption within the UK maritime sector is lagging

compared to other nations making it very difficult to compete.

- Cost is critical to the sector resulting in a reluctance to invest.
- Connectivity presents a major barrier to new and emerging technology – very few industrial sites have WiFi compared to other manufacturing industries where it is standard.

Furthermore, a major factor which is influencing the adoption of I4.0 appears to be a lack of understanding regarding the technology and its benefits. Businesses should take the time to research the different options that are now available and used in other industries. This could be in the form of benchmarking activities, higher education courses, conferences etc. This should be undertaken across all functions and include all levels of employees to ensure engagement across the business. In order to train future generations, it is key that efforts are made to capture as much existing knowledge as possible before the aging workforce retires.

4.2 KEY TECHNOLOGIES

As displayed in Figure 2, whilst the most popular result was automation, there is also a clear realisation that the amount of data available is going to dramatically increase through better connectivity and condition monitoring. There was also clear concern over the security of this data and who could potentially gain access to it. This is likely to cause issues both in term of implementation and effective utilisation of this data.

Figure 2: Survey results obtained on "What technologies do you see having the largest impact on the maritime industry in the UK" question.

Some of the key points of the conversation that followed are presented below:

- It is widely accepted that connectivity is key to the future of the maritime industry, but also that the sector is significantly behind other manufacturing sectors in terms of infrastructure.
- Digitisation is key to helping ensure profitability for the industry by allowing for increased design maturity, risk mitigation and training solutions prior to any physical building operation.
- Adoption of automation is seen as a major focus –There is a perception that the UK is lagging behind with respect to other countries.
- Communication is poor, and businesses rarely collaborate with one another resulting in siloed businesses – many other industries have become much more open to greater collaboration through digital means.
- Additive manufacturing is seen as having an important role to play in particular for retro-fitting and repairing aging vessels.

Besides, data is going to play a major role in the future of the sector, however caution is advised to avoid potential data overload. This could lead to misinterpretation and potentially poor strategic decisions. Businesses should begin setting out a clear and well-defined plan looking at least 5 years into the future to successfully scope what data is required, how it can be accessed, processed and how to present it. A similar approach would also be recommended for implementing automated solutions to assess the potential applications and to highlight benefits. These tasks are not likely to form a part of day to day business for many in the sector, and the job roles involved are often highly skilled and expensive making recruitment difficult. Businesses could also consider subcontracting these skills externally for the early stages of the implementation.

4.3 KEY DRIVERS FOR I4.0 ADOPTION IN THE BUSINESS

As shown in Figure 3, in most businesses, cost is the main reason for looking into new technologies and processes. There is a clear realisation that the sector must adapt to compete with potentially cheaper alternatives around the globe.

Figure 3: Survey results obtained on "What do you believe is the key driver for your business to embrace I4.0 technologies and solutions" question.

Some of the key points of the conversation that followed are presented below:

- Cost management and creating savings within the businesses are key.

- An alternative to producing "savings" is increasing productivity. Enhancing productivity increases profitability without the need to cut back on staff and other expenses.
- Greater adaptability in response to risks and will reduce lead times and increase profitability.
- It was highlighted that providing a quality product/service is not as critical as reducing costs with the business.

It was expected that reducing costs and increasing potential returns would be the key drivers when making business and strategy decisions concerning the implementation of new technologies. However, it is important to look beyond this. Due to the ageing workforce and risk adverse nature of the sector, there is a focus on the "here and now" rather than looking to the future. It is also worth considering other potential benefits that implementing new and innovative solutions could have. Investing in I4.0 technologies and solutions can boost employee moral which can for example, enhance productivity, marketing opportunities, and improve stakeholder perception of the business.

4.4 BARRIERS TO ADOPTING I4.0

As shown in Figure 3, Cost is the major factor considered when implementing I4.0 solutions. The second largest concern revolve around a lack of awareness which is leading to high level of apprehension in the sector.

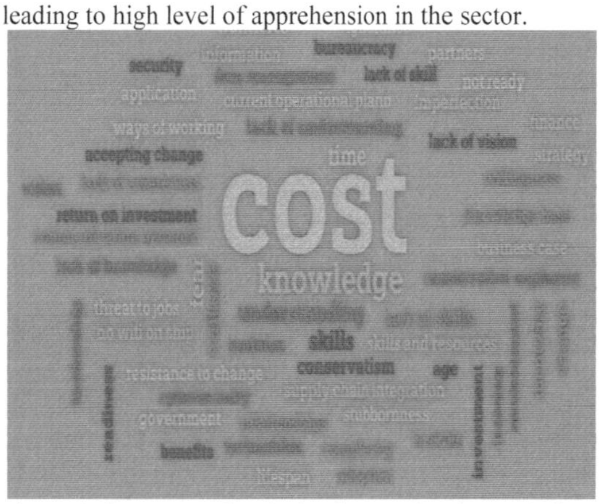

Figure 4: Survey results obtained on "What do you believe are the most significant barriers to I4.0 technologies" question.

Some of the key points of the conversation that followed are presented below:

- Internal cultural change was highlighted by most participants, the aging workforce are resistant to change preferring the traditional paper-based systems while the younger generation prefer digital solutions.
- Implementation projects are typically undertaken in a gradual step by step manner, this means that these projects tend to have long lead times.
- A lack of understanding of the available technology makes carrying out a benefit analysis particularly difficult.
- There is an adversity to standardisation of product offering - while each project will have unique elements, processes by enlarge remain the same.
- There is currently limited work being carried out to plan and establish future processes.

While cost is a major concern and is very difficult to remove as a barrier, there are solutions which could be pursued to help alleviate this. As previously stated, greater education surrounding, and awareness of innovative and disruptive technology and its potential applications could have a dramatic impact. This would allow employees to clearly identify the potential benefits and if combined with accurate quality control data, would allow for a comprehensive benefit/ROI analysis. By promoting the technologies as a method of assisting employees in their work, the fear and apprehension experienced by many of the workforce could be mitigated. Within the marine industry, particularly with the ageing workforce in the shipbuilding sector, there is a fear that some disruptive and innovative technologies, such as co-bots, are being investigated in order to replace the human workforce. This is not necessarily true. Upskilling existing employees and providing high-quality training is a key part of any implementation project to ensure employee engagement.

4.5 CRITICAL I4.0 SKILLS AND ROLES

The results depicted in Figure 5 indicate data analysts will be key in the future of the maritime sector. The results also indicate that the specific types of engineering which will be critical are electrical and software.

Figure 5: Survey results obtained on "What skills or job roles do you view as being most critical in the next 5 years" question.

Some of the key points of the conversation that followed are presented below:

- Within every group, there were discussions around data analysts.
- There was a consensus that the industry has issues retaining graduate engineers in the field of engineering - many are being lost to banking industry.

- It was recognised that some roles may need to be outsourced due to lack of internal skills or that these roles are not required for day-to-day activities.
- Apprenticeships are key in the sector. There should be greater engagement from colleges to upgrade apprenticeship programs in order to align with industry needs.
- It was noted that when considering the future employees of the marine industry, soft skills and personality traits, such as creativeness, ability to problem solve, and drive, are just as important as technical ability.

The industry appears to believe that data analysts will be key to the future of the industry. However, it is important to acknowledge that data analysts are only as valuable as the data they have access to. Therefore, a more critical challenge in the immediate future is ensuring that accurate and useful data can be captured. This could be achieved through changing employee behaviours and well-established processes. While it is extremely likely that the amount of data available to capture is set to increase dramatically within the next 5 years, most of the analysis could be performed by existing enterprise management software. In order to meet the future needs of the sector, it is recommended that the academic institutions and industry work in tandem more efficiently to identify key skill gaps and develop courses that tackle the issues years before they become a prevalent problem.

4.6 INTERNAL AVAILABILITY OF I4.0 SKILLS

The results depicted in Figure 6 are evenly split between the three options. However, the fact that only one third of the respondents felt that they had the sufficient skills in house to embrace I4.0 highlights a potential and significant skill deficit in the near future.

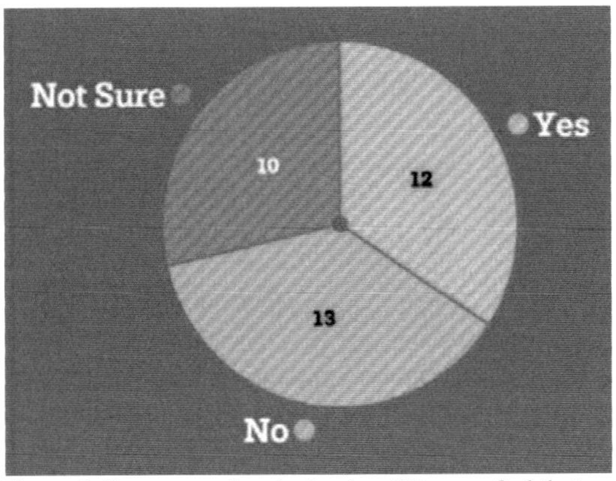

Figure 6: Survey results obtained on "Do you feel that you have the required skills internally to successfully complete your I4.0 journey" question.

Some of the key points of the conversation that followed are presented below:

- It was discovered that leading up to this question, some of the attendee's opinions had changed from "Yes" to "No" due to the day's previous discussions around I4.0 technologies. This indicates that the availability of internal skills is not fully understood or has been assessed yet.
- As referenced previously in 3.1.3, greater communication between academia and industry was raised as a requirement for success.
- There was a level of uncertainty as to what skills would be required as the technology requirements were not fully understood or defined.
- Attracting skills externally could prove difficult due to high competition in the job market.
- Many felt that a combination of internal skills with fresh, external eyes would be required to improve the businesses.

A program of internal capability assessments could be a useful exercise to perform. It is likely that there will be hidden talents and skills internally which are not currently utilised in the employee's role. This assessment would be a valuable resource in the journey towards implementing new technologies and would allow for the creation of a skills gap analysis which would drive the recruitment focus. The general opinion of the attendees was that the required skills could not be identified without first determining the required technologies to be implemented and the end goal to be achieved. This highlights the fact that a clear technology strategy should be created to provide a detailed implementation roadmap.

4.7 METHODS FOR ACQUIRING NEW SKILLS

Following the results presented in Figure 7, the majority of respondents listed Internal Upskilling as their primary method for acquiring the skills for the future of their business. The remaining five sections also received a significant number of votes, suggesting that a number of different methods will need to be pursued.

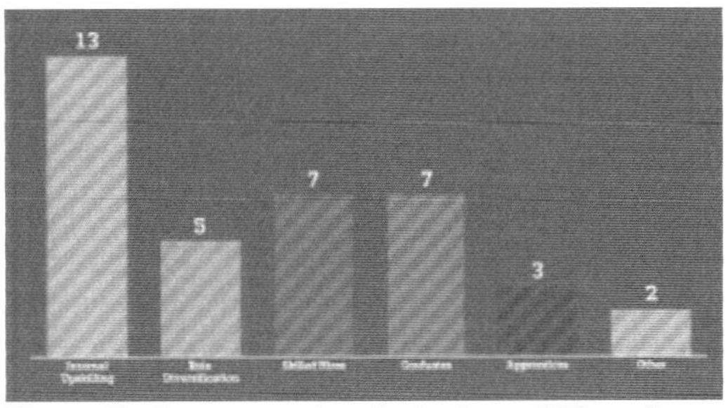

Figure 7: Survey results obtained on "What do you envision being the primary method for acquiring the skills for the future of your business" question.

Some of the key points of the conversation that followed were:

- Expected that a balanced approach will be needed.

- Greater collaboration and potential partnerships could provide the best way forward.
- There was a sense that the industry is not especially good at planning for the future and are instead focus more on the "here and now".
- Feeling that university and college courses must modernise and involve more industry experience.
- There was recognition that apprentices tend to stay with the same business for longer than graduates. It was expressed that apprentices are often more willing to take time to rise through the business whilst graduates can be less patient to do so.
- External hires could be an effective way to change the culture within a company.
- Increased options for apprentices – offer graduate courses upon time served.

The establishment of an industry wide skills network could prove to be an extremely useful resource which would allow for greater collaboration across the sector. This would allow organisations to request personnel, with a specific skill set, from another business on secondment. Whilst in the company the employee would help to upskill existing staff and support the implementation of new technologies. It was also felt that young people are being prioritised for training, however it is recommended that upskilling and development programs are still offered and encouraged to existing employees to avoid disengagement with the workforce.

4.8 FUTURE OF MARITIME IN INDUSTRY IN UK WITHOUT MODERNISATION

There was unanimous agreement that the UK marine sector will require modernisation for the sector to remain viable and globally competitive.

Some key point of the discussion that ensued were:

- While the industry recognises that it needs to modernise, they are unsure when this can and will start.
- It was recognised that modernisation of the full maritime industry must also include the modernisation of the wider supply chain.
- The sector needs strategic initiatives and major action – prevent more and more talking shops.
- Many attendees acknowledged the fact that in order to compete globally, the UK maritime industry needs to be visionary in the sector and to differentiate through innovative technology and processes.
- It was felt that there is a lack of pride in shipbuilding across the UK and that there are not enough young leaders in the industry to effectively change the current culture.
- Customers currently pay premium prices for UK built ships however the quality of production does not warrant the price.
- The technologies and processes associated with the third industrial revolution have not yet been achieved by most of the marine sector and so attempting to embrace I4.0 is premature.

Therefore, for the UK maritime sector to survive and compete globally, it is essential that the modernisation process begins immediately. This is essential in order to avoid falling further behind other nations who have already begun their I4.0 journey. While it is unlikely that dramatic technological changes will be made immediately, the industry can begin to effectively strategise in a collaborative fashion. This will help drive forward investment by providing a clear scope of activity and outlining the future of the industry.

5. CONCLUSIONS

The described in section 4 provided valuable insights into the current state of the UK Maritime sector and informed those present of potential applications of Industry 4.0 technologies. The survey results and follow on round table discussions demonstrated an understanding of Industry 4.0, but a difficulty in how to apply the principles in the Maritime sector. It was widely recognised that Maritime compared to other sectors is falling behind in the uptake of new and innovative technologies. There was also a clear concern over the aging population of the workforce and issues around future skill availability and the appeal of the industry to emerging talent pools.

For the sector to survive it is essential that the industry, government and academic institutions work together in collaboration. If this does not take place, the likelihood of the industry failing and weakening UK manufacturing becomes increasingly more likely. There is clear indication that the government is determined to maintain Naval capabilities, however the same cannot be said for the maritime sector and its wider supply chain. A summary of the common issues identified across the maritime industry is as follows:

- Cost is the largest motivator in the sector preventing the industry from looking at the bigger picture.
- The industry has a significant age gap in the work force with key skills and knowledge soon to be lost.
- Compared to other industries, such as the automotive and aerospace industry, the maritime sector is falling behind in the implementation of new, innovative and disruptive technologies.
- The UK maritime sector is finding it increasingly difficult to compete on the global market in regard to cost and therefore, it needs to compete through innovation
- The industry needs government support to enhance its infrastructure.
- Communication & collaboration is poor across the industry.
- The culture and leadership across the industry must change drastically in order to successfully

implement new technologies, skills and processes.

For the UK Maritime sector to maintain and improve its national and global position, the transition towards Industry 4.0 principles and methods of manufacturing must start now. A clear holistic strategy must be formulated for each company, as well as internal business strategies for the adoption of technology, skills and required cultural changes to ensure the success of implementation. Key aspects that should be considered while designing this strategy are:

- Industry, government and academic institutions must communicate and collaborate more effectively
- The industry must provide a unified voice and lobby the government for support and present a clear, well established strategy to drive investment and the creation of a Maritime Catapult centre.
- The industry must now begin to strategise how it will modernise both internally and as a collective.
- College and university courses need to modernise to meet the current needs of the industry and attract the young skillforce.
- The industry needs to envisage the future of the marine sector and recognise that the initial expenditure for new technology will reap benefits over time.
- The culture and leadership must develop and learn from other industries already on this journey via greater cross industry collaboration and extensive benchmarking activities.

6. ACKNOWLEDGEMENTS

IN 4.0 has received funding by the European Regional Development Fund (ERDF) through the INTERREG Atlantic Area Programme.

7. REFERENCES

Clarkson PLC. (2018). *Annual Report 2018: Smarter decisions. Powered by intelligence.* London, Great Britain.

Inmarsat. (2018). *Industrial IoT on land and at sea - Inmarsat Research Programme 2018.* London, Great Britain.

The Royal Institution of Naval Architects
8-9 Northumberland Street
London, WC2N 5DA
United Kingdom

ISBN 978-1-7138-0869-5

CURRAN ASSOCIATES INC.
proceedings.com